THE CURRENT STATUS AND FUTURE
OF PARASITOLOGY

RECENT MACY PUBLICATIONS

The Current Status and Future of Academic Pediatrics, edited by Elizabeth Purcell

The Psychopathology of Children and Youth: A Cross-Cultural Perspective, edited by Elizabeth F. Purcell

Emergency Medical Services: Measures to Improve Care, edited by John Z. Bowers and Elizabeth F. Purcell

The Current Status and Future of Academic Obstetrics, edited by John Z. Bowers and Elizabeth F. Purcell

Receptors and Human Diseases, edited by Alexander G. Bearn and Purnell W. Choppin

Aspects of the History of Medicine in Latin America, edited by John Z. Bowers and Elizabeth F. Purcell

Protein-Energy Malnutrition in Barbados: The Role of Continuity of Care, by Frank C. Ramsey

The Impact of Health Services on Medical Education: A Global View, edited by John Z. Bowers and Elizabeth F. Purcell

A complete catalogue of books
in print will be sent on request

* * * * *

All publications should be ordered directly from
Independent Publishers Group
One Pleasant Avenue
Port Washington, New York 11050

Edited by
Kenneth S. Warren
and
Elizabeth F. Purcell

THE CURRENT STATUS AND FUTURE OF PARASITOLOGY

Report of a Conference
Sponsored Jointly by
The Rockefeller Foundation
and the
Josiah Macy, Jr. Foundation

JOSIAH MACY, JR. FOUNDATION
One Rockefeller Plaza, New York, New York 10020

© 1981 Josiah Macy, Jr. Foundation
LIBRARY OF CONGRESS CATALOG CARD No.: 81-84673
  Warren, Kenneth S. and Elizabeth F. Purcell
    The Current Status and Future of Parasitology
  New York: Josiah Macy, Jr. Foundation
  296 pp.
  8110   810929
ISBN: 0-914362-37-2
Manufactured by the Heffernan Press, Worcester, Massachusetts
Distributed by the Independent Publishers Group
One Pleasant Avenue, Port Washington, New York 11050

DEDICATION

To John Zimmerman Bowers, M.D.

A cynosure of medical philanthropy

CONTENTS

FOREWORD

Parasitology is a branch of zoology with particular significance for the well-being of both mankind and domestic animals. The parasitic diseases are a major component of the Rockefeller Foundation's program of biomedical research on "The Great Neglected Diseases of the Developing World." They comprise a large proportion of the subject matter of the clinical field of tropical medicine concerning which the Josiah Macy, Jr. Foundation convened a seminal conference eight years ago.*

When John Z. Bowers of the Macy Foundation suggested a follow-up meeting on tropical medicine some time ago, I raised the question of holding a symposium on the current status and future of parasitology. Bowers was enthusiastic about the idea, and John A. Pino, director of Agricultural Sciences at the Rockefeller Foundation joined us with alacrity. It seemed to all of us that the field of parasitology was suffering from many decades of neglect, both by funding agencies and by the scientific establishment. This in spite of the fact that these major diseases of known causation offer a unique opportunity for research, particularly in view of the plethora of new scientific tools of biology, immunology, and chemistry.

We decided to look at a series of aspects of parasitology: history, teaching, research, training, career opportunities, funding, the literature, and the future of the field as exemplified by the scientific opportunities afforded by new developments in cell biology, molecular biology, pharmacology, biochemistry, immunology, and ecology.

The participants in the conference, held on 20 to 22 Oc-

---

* "Proceedings of the Macy Conference on Teaching Tropical Medicine," *American Journal of Tropical Medicine and Hygiene* 23 (suppl.) (July 1974).

tober 1980, included great classical parasitologists as well as a group of scientists recently attracted by the challenge of this neglected but highly accessible field of biology. The chemistry of the meeting was enhanced by the manifold charms of New Orleans, and a unique melding of the best of the classical and the modern biologies occurred that could lead to a veritable renaissance of parasitology. The benefits of such a rebirth could have enormous implications for the health of the people and domestic animals of this world.

Kenneth S. Warren, M.D.
Director, Health Sciences
The Rockefeller Foundation

1 July 1981

WHAT IS PARASITOLOGY?

Paul C. Beaver

I have been unable to find an acceptable brief definition of parasitology; nor have I been able to construct one. Parasitology is the scientific study of parasites, but this leaves us with the task of defining the terms *parasite* and *parasitism*. Here is where difficulties arise. As the theme of this conference is the present status and future of parasitology, however, it is reasonable that we begin by examining the nature and dimensions of the subject—its boundaries and major areas of present and future interest.

I expect that all parasitologists on entering the field are astonished to learn that parasites exist in such a wide range of species, and that nearly all kinds of animals are natural hosts to nearly all kinds of parasites. I read a recent statement that there are now about 65,000 known species of protozoa, 10,000 of which are parasitic.[1] In that instance the term parasitic must of course include species that might be called *commensal* or *symbiotic*. If organisms are viewed as either free-living or parasitic there is hardly an animal in nature or a person in any part of the world who escapes serving as host to animal parasites.

Here I should mention that while it is appropriate to classify bacteria and fungi as either free-living or parasitic—and viruses too are parasitic—these forms generally are not included among the elements of parasitology. Thus it can be said that parasitology is the scientific study of parasitic protozoa,

helminths, certain lesser groups of worms, parasitic arthropods, and the vectors of parasites, that is, parasitology is
largely an amalgamation of protozoology, helminthology, entomology, and acarology; but of course all these disciplines
cover free-living as well as parasitic species.

This introduces the question of how we should classify
free-living organisms such as *Naegleria, Acanthamoeba, Micro-
nema,* and others that sometimes invade and colonize the
tissues, but are not transmitted from host to host; in this relationship the colony is parasitic, but the species is not. I mention
them here because a knowledge of these opportunistic parasites, or at least an awareness of them, can be useful in studies
of the relationships between true parasites and their hosts.
What habitat conditions, we may ask, does the brain of a human
or of a horse or mouse provide—or not provide—to which
these amphizoic invading organisms are adapted in their
natural habitats in water or soil? And what facts can be learned
from them that contribute to a better understanding of
parasitism?

It need only be mentioned that parasitology extends
throughout the biological world, including hosts of all
taxonomic groups and parasites that represent many different kinds of protozoa, helminths, and arthropods. Parasitology
encompasses, rests on, and draws from essentially all branches
of biology, including morphology, taxonomy, ecology,
epidemiology, physiology, pathology, immunology, biochemistry, nutrition, genetics, sex and mating behavior, molecular
biology, aquatic biology, and soil science. Also, there are
parasitologists whose primary orientation is to medical sciences,
public health, animal husbandry, and toxocology.

Parasites are of interest to the naturalist biologist and of
great importance to the medical scientist and practitioner. To
the one, parasites are a part of nature, to the others they are
perhaps natural in frogs and fish, but unnatural in man and to
a certain extent in domestic animals. To the biologist, intestinal
nematodes are viewed in the same light as root or soil
nematodes, or, for that matter, meadow mice, wood rats, or
arctic foxes. The medical and veterinary practitioner, on the
other hand, must view them from the standpoint of health and

disease and be prepared to assess them from the point of view of the patient.

Patients generally are eager to be rid of their parasites even when they are essentially harmless—or even when they are nonexistent. A parasitologist not only must deal with all the various kinds of pathogenic and nonpathogenic parasitic animals, but must expect occasionally to encounter people whose most troublesome parasites are those that exist only *in the mind*. For those who see patients, therefore, parasitology must include problems presented by people with delusions of parasitosis, who firmly believe their skin is invaded by crawling, biting, flesh-consuming parasites.[2] In such cases the diagnosis usually can be made quickly, but without a knowledge of parasites a hasty diagnosis could be wrong.

A few weeks ago I was shown material from a recent case of *Sparganum proliferum,* a condition in which small larval cestodes actually do invade and multiply in the skin, spread to most parts of the body except the face, and constantly, one after the other at intervals of one to several days, are seen emerging from the skin.

Hosts and parasites, having evolved together, have found ways to live together with mutual tolerance under natural conditions. To some degree certain human populations have developed an appreciable tolerance of certain parasites, as, for example, the malarias and hookworms, but in general the parasites' natural adaptations for sparing the host probably have contributed more to this mutual tolerance than have any specific resistance mechanisms in the host. In general the hosts could have survived and evolved without parasites, but the parasites could not exist as such without hosts.

Usually any attempt to formulate a precise definition of parasitism rests heavily on poorly founded assumptions about the parasites' nutrition and on the notion that a true parasite must be harmful. Some workers stipulate that the ability of the parasite to kill the host differentiates parasitism from commensalism.[3] I mention this to raise the question of whether the immune relationship between host and parasite differs basically from that between host and commensal. There can be no doubt that one of the most basic aspects of parasitism is population control, that is, of the parasite population, not of the host. The parasite's role in limiting a natural host population apparently is relatively minor—and exceptional. From an evolutionary point of view it would seem highly advantageous to the parasite to endow itself with the capability to induce a favorable host response, or no response, rather than an unfavorable response. Population control ought to be among the chief characteristics of a parasitic species. This would be equally as true for colonizing parasites such as the blood protozoa and itch mites as for the noncolonizing helminths such as hookworms and liver flukes.

The natural and common existence of a host-parasite relationship in which the presence of a well-tolerated population is protective of both host and parasite against the acquisition of a burdensome increase in the parasite population was once a widely accepted view and was discussed by biologist-parasitologists under the term *premunition*. The term is seldom used now, but it may be essentially what is called concomitant immunity. In any case the immune relationship is a factor to be considered in the investigation of any aspect of parasitology, except perhaps pure morphology.

About forty years ago Justin Andrews wrote two articles which, at that time, caused considerable dissension among physicians and public health workers.[4,5] The main points emphasized by Andrews were that most hookworm infections in endemic areas are light, and therefore subclinical, and that the primary objective of public health authorities should be the detection, prevention, and control of hookworm *disease* rather than the elimination of subclinical hookworm *infection*. He also pointed out, indirectly, that when organized efforts are focused on a single major parasitic disease in a particular region or community, such as hookworm disease in Georgia, the illnesses resulting from other causes may be mistakenly attributed to the target parasite.

Parasite population factors must be among the major concerns of parasitology. It is obvious that in areas of high prevalence the parasite burden in most infected individuals does not represent all, or even more than a mere fraction, of the total inoculum. From this observation two questions arise: 1) What are the natural factors that normally restrict the parasite populations in most infected people?; and 2) What are the abnormal or unnatural factors that permit or cause overpopulation of the parasite in a few? That is, what are the factors that permit or cause the heavy disease-producing infections? One of the concerns of parasitology, therefore, is the quantitative assessment of parasite burden and the establishment of thresholds of disease-producing levels of infection.

An axiom of helminthology, not universally accepted, but generally true, is that, although a few worms of any species will rarely be perceptibly harmful, all species of worms are harmful when present in massive numbers. As already noted, however, under natural conditions massive infections are exceptional. Applied to human populations, the conditions that favor the acquisition of damaging infections are found in periods and places of transition, from the primitive, where populations are small and scattered, to the highly developed, where although large and densely settled the populations are comfortably housed, well-fed, and well-served with sewerage systems. In this way community development becomes one of the concerns of parasitology.

Many of the parasites of domestic animals have these same problems of population control. Animals or people assembled in large numbers and confined to small areas for long periods create special problems that may fall to the parasitologist to solve. Well-known examples taken from hookworm history are the tea workers of South India, Sri Lanka, and Malaysia; tunnel construction workers of Europe; and underground mine workers of Europe and of this country some years ago.

In summary, parasitology can be defined as a study of the relationship between parasites and hosts, parasites being, generally speaking, animals that live in or on the hosts. Parasites are dependent on their hosts and therefore have developed, by adaptation, ways of maintaining under natural circumstances tolerable conditions for living and reproducing in their hosts. The principal aim in parasitology is to acquire sufficient knowledge about parasitism in natural populations to control or eliminate parasites from unnatural populations, both of the people and of the animals on which the people depend.

In the introduction to Part II of his famous monograph on the hookworm *Ancyclostoma duodenale*, Arthur Looss stated what all workers in parasitology can usefully keep in mind:

> Thus it happens that not infrequently a theory, perhaps very clever and acute, as to the presumable development of a worm is built up . . . , and finds acceptance and dissemination, although it is simply impossible helminthologically, because it attributes to the parasites things which they can never do. We often do not know the special behavior and special development of a certain intestinal worm of Man, but if the study of related forms living in animals were undertaken we should attain a fair certainty as to what the truth will be. The familiarity with the parasites of animals enables the observer, in many cases, to say at once whether a theory set up with regard to a parasite of Man is intrinsically probable, i.e., whether or not it can be correct. . . . What I wish here to emphasize is that a correct knowledge of the diseases of man caused by worms, and all that is connected with them is the more difficult to attain the more the parasites of animals are ignored.[6]

NOTES

1. N. D. Levine, J. O. Corliss, F. E. Cox, et al., "A Newly Revised Classification of the Protozoa," *Journal of Protozoology* 27 (1980): 37–58.

2. "Delusions of Parasitosis," *British Medical Journal* (26 March 1977): 790. See also "Letters to the Editor," (7 May 1977): 1219.

3. H. D. Crofton, "A Quantitative Approach to Parasitism," *Parasitology* 62 (1971): 179–93.

4. J. Andrews, "New Methods of Hookworm Disease Investigation and Control," *American Journal of Public Health* 32 (1942): 282–88.

5. _____, "Modern Views on the Treatment and Prevention of Hookworm Disease," *Annals of Internal Medicine* 17 (1942): 891–901.

6. A. Looss, *The Anatomy and Life History of Ancyelostoma duodenale Dub.*, vol. 4, pt. 2, *The Development in the Free State* (Cairo, Egypt: The University of Cairo, 1911): 175.

DISCUSSANT:

Kenneth S. Warren

Parasitology is a subdiscipline of biology that comprises as its subject matter protozoa, helminths, and arthropods; the last are mainly insects that may directly cause disease or, more important, transmit protozoan and helminth infections. The practical importance of these organisms to the well-being of man and domestic animals is enormous, and of particular interest to veterinarians and to physicians in the field of tropical medicine.

Parasitic diseases are among the major problems in the developing countries, which are largely in the tropics and account for most of the world's population. Nevertheless the parasitoses have recently been called the "great neglected diseases of mankind." There are many reasons for this, including the lack of a scientific infrastructure for studying them where they are most abundant; the relative lack of interest of the great scientific establishment of the developed world; and even the belief that the health of the people of the less-developed countries is an unimportant factor in the advancement of those areas—see, for example, the famous Pearson report.[1]

Fortunately, many of these situations and ideas are beginning to undergo changes. That is the reason the Macy and Rockefeller foundations have convened this conference on the "Present Status and Future of Parasitology." It is doubtful that anyone present, no matter of what persuasion, believes that the field of parasitology is in a healthy state, enjoying the well-being of more than adequate funding, attracting the finest students, and functioning at the forefront of modern scientific methodology and ideas.

Concerning the excellent disquisition, "What Is Parasitology?" by Beaver, one of its greatest practitioners, I would like to raise the question of whether there may be something inherently wrong with the discipline as it has evolved and as it is practiced that may contribute in some small but significant way to its present status.

It is obvious that every scientific field has its "insiders" and its "outsiders." The insiders are those who have been trained through their doctorates as parasitologists; the outsiders have doctorates in other fields such as immunology, biochemistry, or medicine. Thus, in an introduction to a workshop on the "Immunology of Parasitic Infections,"[2] the terms immunoparasitologist and parasitoimmunologist were coined; two individuals whose chapters appear in this volume, Dean Befus and Alan Sher, exemplify these respective titles. Although not steeped in the lore of parasitology, and certainly without the breadth and depth of someone trained in the field, it might be agreed that the "outsider" may have much to offer it, may even grow to love it, may have great feelings of loyalty to it, and may attempt to foster it. One advantage of outsiders is that they have different perspectives, and I should now like to take this opportunity to state a controversial position that I hope will lead to a lively and constructive discussion.

It is my belief that the discipline of parasitology as it is now constituted suffers under the pall of a misnomer. One of the few textbooks that have truly recognized this is T. W. M. Cameron's *Parasites and Parasitism*,[3] working from the belief that all infectious agents are parasites. Thus it is my belief that departments of microbiology with their own narrow subject areas, essentially viruses and bacteria (fungi seem to have fallen into a crack), should be renamed departments of parasitology and should deal with all infectious agents including protozoa and helminths. A great university recently considered incorporating its department concerned with tropical health—largely parasitology—with its department of microbiology. The converse might be recommended, however, as the two greatest causes of infant and child mortality in the developing world are the bacterial and viral diarrheas and pneumonias; a reasonable balance might be maintained, since the parasitic diseases are of such overwhelming importance in the tropics. This balance would almost certainly be lost in the opposite situation. In departments of biology the basic phenomenon of parasitism and the fascinating host-parasite relationships involving all such organisms might constitute an exceedingly attractive field. Indeed, Lewis Thomas's fascinating essays often discuss these problems in their broadest sense: an example is "The Medusa and the Snail," the title essay in his most recent collection.[4] Subfields would then consist of protozoology, helminthology, mycology, virology, and bacteriology.

This might alleviate another problem of parasitology: its failure of specialization. The disciplines of bacteriology and virology, though interrelated, usually have separate practitioners. This is reasonable, as the technologies involved are usually very different. As I wrote earlier: "There is a similar dichotomy in the subdisciplines of parasitology and this is dramatized by the relative size of the organisms."[5] The smallest adult multicellular helminth of man, *Strongyloides*, is visible to the naked eye; in contrast, many of the unicellular protozoa are invisible intracellular parasites. Furthermore, the helminths have a host-parasite relationship unique among all parasitic agents, one in which the organisms within the definitive host do not directly replicate. The technologies of helminthology and protozoology, again, are totally different, as are the consequences of the infections and the means by which they are controlled. To enter these fields in proper depth it might be well to attack them as relatively separate subdisciplines of parasitology.

A unifying concept for the entire field of parasitism might lie in the unique host-parasite relationship that involves membranes; localization mechanisms such as lectins and pheromones; toxins; host and parasite genetics; natural, non-specific and specific immunity; and even the newer concepts of mathematical ecology.

Parasitologists have a deep concern about a proper and broad understanding of the taxonomy of the organisms, their complex life cycles, and their cultivation and maintenance both in vitro and in vivo. These important aspects of parasitology must be preserved. But it is also of importance that shirts, trousers, and shoes be shed and that parasitologists dive headlong into the modern biological revolution. This is best exemplified by the new ten-week course in "The Biology of Parasitism," suggested by Joshua Lederberg, president of the Rockefeller University, and directed by John R. David of Harvard Medical School with the assistance of Eli Chernin. The course began in the summer of 1980 at the Marine Biological Laboratory in Woods Hole and will continue for the next four years, training almost 100 people. The course content has been summarized by David as follows:

> The sixteen students were divided into four groups. Experiments, including many which had not been done before, were carried out during the ten-week lab course. Among these, the students produced monoclonal antibodies to *Leishmania enrietti* and demonstrated that several reacted with tubulin; carried out studies on recombinant DNA with leishmania in an attempt to obtain clones of bacteria which express leishmania antigens; isolated messenger RNA from several protozoa and translated parasite proteins with these; demonstrated the presence of the tubulin gene in DNA from trypanosomes and leishmania, but not *Entamoeba histolytica*; demonstrated immunity to schistosomiasis with irradiated cercariae, and learned to quantitate lung schistosomula and perfuse for adult worms; demonstrated the killing of schistosomula by antibody and eosinophils and obtained eosinophil preparations of 95 percent purity; labeled membrane components of plasmodia, trypanosomes, leishmania, ameba, and filaria with a number of surface probes and analyzed the products by gels and autoradiography (one malaria gamete antigen so labeled was isolated with a monoclonal antibody); studied surface parasite antigens using a fluorescence-activated cell sorter; studied the lectin binding characteristics of filariae, eggs from *Schistosoma mansoni* and *hematobium*, and found specific lectin binding patterns for each; demonstrated the presence of a natural ionophore in extracts of *Entamoeba histolytica*, which may play a role in cytotoxicity; demonstrated antigenic variation with *Trypanosoma brucei* and isolated surface variant antigens; demonstrated the presence of several protozoan enzymes that have substrate specificity that differs from their mammalian counterpart; developed a solid-phase radioimmunoassay for antibodies against *Entamoeba histolytica* and demonstrated antigen capping with this protozoan; followed the complete life cycles of mosquitos and snails, plasmodia and schistosomes; and studied the physiology of exflagellation of malaria gametocytes and the immunology of antigamete antibodies.*

---

* John R. David, 1980: personal communication.

As David concluded, "It was quite a summer!"

NOTES

1. *Partners in Development*, Report of the Commission on International Development, Lester B. Pearson, Chairman (New York: Praeger, 1969): 399.

2. K. S. Warren, "Introductory Remarks," in "Immunology of Parasitic Infections: Report of a Workshop," ed. K. S. Warren, J. David, F. A. Neva, et al., *American Journal of Tropical Medicine and Hygiene* 26 (1971): 6–8.

3. T. W. M. Cameron, *Parasites and Parasitism* (New York: John Wiley & Sons, 1956).

4. L. Thomas, *The Medusa and the Snail* (New York: Viking Press, 1979).

5. Warren, "Remarks" (see note 2).

DISCUSSION

CHERNIN: Warren started by saying that parasitology is a misnomer, in the sense in which Beaver used the term, and went on to speculate about combining departments of microbiology and departments of tropical public health and parasitology. Well, the Department of Microbiology I know best has little or nothing in common with the Department of Tropical Public Health.

As to the new MBL course, which I helped John David organize and teach in 1980, I came away feeling it is just as important to know how to do a proper stool examination as it is to know how to produce a monoclonal antibody.

WARREN: I want to make one point very clear: I did not state that one area was better than the other. The purpose of the course at Woods Hole was not to teach students to do stool examinations or to identify mosquitoes. If a course was established for the latter, I would hope it would be as good and as stimulating as the former.

What bothers me about Chernin's remark is that the emphasis on the importance of stool examinations and identification of parasites has had a negative influence with respect to attracting people to the field. We need both stool examinations and hybridomas. Biological research is a continuum, and all aspects of it are important.

HILL: At the course at Woods Hole one postdoctoral student enrolled, Barbara Johnson from Colorado State University, found the course stimulating in terms of several points Warren made. Her previous training was in biochemistry, and she went to the course to try to bridge the gap between what she could gain from parasitology and her excellent background in biochemistry. What she came back with was just fantastic.

MAY: I would like to comment as an outsider.

On the one hand, I believe the individual pieces in the transmission cycle, whether they be a virus, bacteria, protozoa, helminth, or arthropod, are so different in the different disciplines that you couldn't put them all together in a common framework if you wanted to.

On the other hand, when it comes to looking at how rationally to plan an overall public health strategy, one has to look at the overall population biology of the transmission cycle. There, for reasons that are purely historical as far as I can make out, the discipline is in a much less sophisticated state than other areas of ecology, and in the overall life-cycle picture there are more similarities than dissimilarities. You can look in a common framework for the characteristic differences of viruses, bacteria, protozoa, and helminths.

Some of the unifying concepts do that. If you ask quantitative questions about what sort of vaccination schedule should be mounted against malaria, hookworm, or measles, across the spectrum, there are commonalities that do not seem to be appreciated as widely as they might be.

I will elaborate my own vision later, which is why I wanted to put in a preliminary advertisement now. I will talk about some common intellectual framework to do with the overall ecology of host-parasite associations, which has its applications in practical things like control measures, which, within that framework, branch out into the particularities that I see as very different across the spectrum from one to the other.

TRAGER: I agree with Warren that there are generalities of parasitism we can study as a very important and interesting aspect of biology, and that knowing how to do a stool examination is not necessarily relevant to this general study. On the one hand, it is of course relevant to diagnostic parasitology, and it is certainly relevant if studies are being conducted on amoeba or some other intestinal parasite. On the other hand, if you are working with malaria, where you can generally live in a much cleaner way, you would have to know how to recognize a malaria parasite.

The people who want to do molecular biology of parasitism obviously have to learn the basic biology of the organism they are going to work with. This was and should be one of the objectives of a course such as the one given at Woods Hole.

I would like to think, however, that in the study of parasitism we could to some extent look at other kinds of host-parasite associations. I think, for example, of a postdoctoral fellow I had who was working on parasites of a fungus that invaded the host and there, in about twenty-four hours or so, multiplied extensively to produce a crop of zoospores that then came out and invaded the hyphae. There are many analogies with the development of malaria parasites in erythrocytes. We can look for things such as receptor molecules, which would probably turn out to be very different, but there might be some general principles that would be of fundamental interest.

KREIER: I am in sympathy with the view that parasitology should include all the disciplines of microbiology, including bacteriology and virology. As a member of a microbiology department I have a very entertaining time when on occasion I try to convince my colleagues they are all parasitologists.

I have always had the bad habit of creating structures in my mind to describe how I believe the world ought to be. Occasionally I am disturbed by how little the world conforms to some of the patterns I have thought appropriate for it. Custom, determined by historical accident, has shaped our associations almost as much as logic; time and habit make the associations seem axiomatic.

If you are trying to put departments together to give parasitologists a happy home, you probably have to look at each particular case to see whether the individuals involved actually will fit together and be tolerant of and sympathetic to each other.

It is a historical accident that those working with microorganisms developed the science of immunology. But people are used to immunologists in microbiology departments and continue to live with this arrangement. Parasitology, by another series of accidents, developed in zoology departments. It is not easy to change habits.

Helminthologists, particularly, and to a lesser extent protozoologists, did not bother to develop modern biochemical concepts because they could always draw pictures of the complex organisms with which they worked and identify them by morphology.

The microbiologists were confronted with the awful reality that they had only about four different shapes and some size differences, so they had to find something else to use for identification: they invented biochemical means. The biochemistry was simple in the early days. One could put the organisms in cultures containing, for example, different sugars, and use the reactions observed as a basis for classification. The entire modern biochemical classification system was developed because if you worked with bacteria you had no alternative. The parasitologists, on the other hand, continued to base their taxonomy on the rich morphological information available to them.

I agree that the parasitologists ought to be together with the microbiologists—and certainly the techniques of modern microbiology, particularly the biochemical and molecular ones, are immensely valuable to parasitologists—but it may take a little while for them to get accustomed to the idea.

CHERNIN: The business of ossification of departments and what gets hybridized with what can sometimes take on peculiar proportions. I have argued at faculty meetings, for example, that there is no reason why the School of Public Health should not be absorbed by the Harvard Medical School as one department in its overall system, rather than the other way around. I bring this up to illustrate how far the argument can be carried.

SIMPSON: I think we are avoiding the main issue. I agree

with Warren's original statement, but I would like to be more iconoclastic and say that parasitology doesn't have a future because the future lies in problems, not in "ologies." The organism should be respected because it presents a problem, such as how cells or organisms divide, reproduce, differentiate, and die. These are the things we want to find out; whether it is a particular organism, free-living or parasitic, is not important.

At schools such as the University of California, Los Angeles, where I am, the name of the department does not necessarily say anything about the people in it or what they do. The Department of Microbiology has individuals who work with parasites, viruses, and eukaryotes, as does the Department of Biology. Compartmentalization de facto is breaking down and will do so even more in the future.

HILL: I have a question for both Beaver and Warren. Should there be continued emphasis on training individuals in tropical countries when it comes to the type of techniques we are talking about, particularly in the MBL course? For example, shouldn't individuals from developing countries be trained as well as Americans?

BOWERS: The major problem I encountered, beginning in 1952 when I went to India for the Ford Foundation, was that people who came to the United States to train became superspecialists. Thus the impact of American medical science was rather destructive of what we might call general medical care in the developing world.

WARREN: Both positions are valid. We have to get people working on problems of the kind John David described in the MBL course because they are a very important and neglected part of parasitology. But we have to be very careful about training people in these techniques who go back to their own countries and try to continue to use them.

A good example is the fluorescent cell sorter, which David was able to persuade someone to lend him. It is difficult to work with and maintain, and it costs about $350,000. While it is an exceedingly powerful tool, it has to be used in the right place at the right time.

Two of the Rockefeller Foundation's Career Development Fellows exemplify the situation. Anil Jayawardena, when questioned about doing research back home in Sri Lanka, replied, "I am interested in malaria, a major problem of my people. But the level at which I am studying malaria here, I can't do back in Sri Lanka. I feel I am giving something to my people by studying malaria in this country." He was very honest about it, and it is certainly a valid argument.

I am very impressed with the fact that Nadia Nogueira, a Brazilian from Bahia, is working at the Rockefeller University until it is possible for her to go back and work full time in Brazil. She is doing some very important research and has begun to spend a certain amount of time every year in Bahia, combining the human material available there with the laboratory facilities at Rockefeller. Maybe she would like to comment on the issue.

NOGUEIRA: You have put me in a tough spot. The brain drain is a reality all over the Third World. Once you come here and reach a certain level of technology it is very satisfying, and it is hard to go back to a situation where you cannot be so productive. At the same time I don't think that should be used as justification for not trying to introduce new technology and new advances to the Third World, or for not bringing people here for training.

I don't want to generalize from my own experience. Personal as well as professional reasons influenced my decision to stay here. I feel an obligation to my country. I have been working on a disease found in Brazil, and I am trying to give back some of the expertise I have acquired.

But, independent of my own situation, I believe people should be brought here for training. The big question is: How long should they stay? There is a limit beyond which it is very hard to go back. I think a reasonable and important result could be achieved by bringing people to this country for short periods of time.

The main contribution people could make in terms of introducing new developments to the Third World is through communication and exchanges. People can come here and learn new techniques and start slowly reproducing them at home. But the main problem is the scientific isolation people feel most of the time.

BOWERS: I think we should now shift to agriculure and let John Pino make a few comments.

PINO: I am glad we are having this discussion about definitions now, because I hope we will set it aside and not let it become an obstacle to what we are hoping to accomplish with this conference.

As far as I am concerned, parasites are the enemy. With due respect, Dr. Beaver, when they reach populations I don't think they know they should not overwhelm their hosts, but the balance isn't always so nicely struck. I believe this is true of any of the organisms with which we are concerned in agriculture.

We held a meeting at the Rockefeller Foundation offices a few years ago when we talked about similarities of and differences between plant and animal and human diseases. Can you imagine a group of plant pathologists talking to human disease specialists and immunologists? Yet when they talked about basic molecular mechanisms, they felt they had some things in common.

And so my plea is that we take the science and the technology available in whatever field, whether it is dealing with viruses, bacteria, plant disease, or anything else that can be useful in understanding the behavior of most parasites and the hosts they invade.

BOWERS: When I was dean of medicine at Wisconsin in 1955, the president, E. B. Fred, who had done a great deal of work on plant pathology, told me that the first two American colleges of agriculture to do fundamental research were Cornell and Wisconsin. I think Wisconsin had the first Department of Genetics, as well as a Department of Plant Pathology and Biochemistry. Would you comment on the development of the colleges of agriculture as scientific institutions.

PINO: I'm not sure I am qualified to talk about the evolution of research in colleges of agriculture, but I don't think there is anything different about their capacity—and I am including veterinary medicine in that general classification—to do fundamental research than a school of biological sciences or a school of chemistry.

Unfortunately there is little interaction among the various groups that could be helpful to one another. And so they go on independently. That may be a luxury that more and more we can't afford in this country or in many other countries around the world.

The colleges of agriculture, in terms of plant improvement generally, which to me goes from genetic improvement to protection against disease, and so forth, are doing excellent work today, which at the molecular level is equivalent to what is going on in many other areas.

It is time for us to make it possible for people to have a place where they can interact about some of these phenomena.

WARREN: Agriculture has been my model since I joined the foundation. I am impressed with what agriculture is doing. It encompasses a vast series of factors—hunger and the fact that people are starving, at one end, and food production at the other, with an enormous number of factors in between.

What agriculture sets out to do is well-exemplified in the title of its program "Conquest of Hunger." Dealing with only one end of the spectrum the agronomists and geneticists set out to produce more food and they were enormously successful. They were buying time for us in the world today, with our population problems, simply because they set out to produce more food and did not try to do everything on the spectrum.

The model for our health program is not institution-strengthening in developing countries, which is important and is being fostered by the World Health Organization, but finding out something crucial about *T. cruzi,* for example, that will enable us to prevent Chagas' disease.

Our goal should be to mobilize the scientists of the world wherever they are to help us develop means of disease control. We are not concerned with elite education in the tropics in this area, but with the best ways to develop effective vaccines and drugs and applying them.

One of the most sophisticated biochemists, who is very concerned with the application of what he is doing, is Anthony Cerami, a graduate of an agricultural college.

CERAMI: As a biochemist who was trained at a college of agriculture I understand what you are saying about models and how agriculture has had a dramatic effect on American society. What it did in the early part of the century was to take problems of food production and apply science to them. The advances made were incredible.

But one thing that happened—and we may argue about this—is that, from about the 1950s on, agriculture in the United States became ossified, except for plant genetics. Certainly it is true of animal husbandry and animal disease; the veterinary schools have not progressed at the same rate as schools of medicine.

In the 1940s there were superior departments of bio-chemistry and microbiology in colleges of agriculture. Strep-tomycin, as an example, was discovered in the Department of Soil Microbiology at the Rutgers College of Agriculture. Today the colleges are suffering, and that is why the USDA has broadened its whole concept of giving grants to institutions other than land-grant colleges and regional places. So agricul-ture is a model to follow to a degree, but one needs to be careful not to follow it too far.

WEINSTEIN: This discussion of agriculture is critical, for it points to the intellectual fragmentation that has been going on because of the creation of departments within institutions, which has resulted in the separation of "disciplines." We see this exemplified by the very nature of the organization of this col-loquium on the future of parasitology.

We have been discussing the importance to the world of agriculture and food production, yet we do not have the par-ticipation of a plant nematologist who could have discussed the objectives and trends in research on this vast group of parasites. In the past several decades it has become increasingly evident that plant parasitic nematodes have a profound influence on food production. Unfortunately a schism has developed in the study of nematodes such that little direct contact and intellec-tual discourse occurs between free-living animal and plant nematologists.

Fundamental research on the biology of nematodes is being conducted by all these individuals, yet they usually stand cold shoulder to cold shoulder. Discussions of "the future of parasitology" should take into account ways to restore a com-monality and unity of scientific purpose among such divergent groups. I believe it would lead ultimately to a truer understanding of the principles that underly the phenomenon of parasitism.

In the same way, I would argue against the suggestion presented earlier that our study of parasites should be predicated on whether they are protozoa and differ in size from helminths, or whether helminths as parasites differ from protozoa in that they do not multiply within a host—that is of relatively inconsequential significance.

WARREN: Would you include bacteria and viruses?

WEINSTEIN: In the framework of modern scientific life, probably not, because we are unfortunately dealing with the ossification of a historical process of departmentalization. In terms of what we are discussing here, I believe we should focus more on the unity of parasitism and the phenomena that are expressed from the biochemical to population levels.

A BRIEF HISTORY OF
AMERICAN PARASITOLOGY:
THE VETERINARY CONNECTION
BETWEEN MEDICINE AND ZOOLOGY

Calvin W. Schwabe

In highlighting the history of American parasitology I shall adhere strictly to our initial conference charge by indicating some of the ways the "integration of medicine and veterinary science" has furthered the development of the field. Because many human parasitisms are zoonoses or involve animal vectors, because much human medical research can be carried out only in animal subjects, and because veterinary pathology is merely another way to say abnormal zoology it was inevitable that work on animal parasites and the diseases they cause should have been one of the first—and have remained one of the most active—areas of common interest and cross-professional efforts among physicians, veterinarians, and zoologists.

It is therefore not surprising that very important contributions to the development of human parasitology/tropical medicine in many countries should have been made by zoologists and veterinarians.[1-3] Of the latter, among national pioneers in these fields whose names come quickly to mind are Peter Abildgaard in Denmark, Karl Rudolphi in Germany, Eduardo Perroncito in Italy, Alcide Railliet in France, Sir Arnold Theiler in South Africa, T. W. M. Cameron in Canada, Marcos Tubangui in the Philippines, G. G. Witenberg in Israel, Robert Daub-

21

ney in England and Kenya, Konstantine I. Skriabin in the Soviet Union, and Ian Clunies Ross in Australia. That similar input by animal biologists was also true for the United States is clearly reflected in the statistic cited by H. E. Meleney and W. W. Frye in 1955: only 34 percent of teachers in these disciplines in American medical schools then possessed the M.D. degree.[4]

I think most of us here would regard these past relationships as desirable, for it seems beyond question that mutually refreshing interactions among scientists with complementary as well as shared interests, paradigms, knowledge, and skills cannot help but work to the advantage both of medicine and the public. It is therefore a matter of importance to the future of human parasitology/tropical medicine in America that these productive precedents for cooperation remain clearly understood and continue to be cultivated actively.

I might add parenthetically that, as a zoologist and veterinarian whose associations with schools of human medicine, veterinary medicine, and public health have been equally intensive and rewarding, I have for a long time emphasized the benefits to be realized from more cross-professional efforts in medicine generally. For in my opinion the fruitful interactions from parasitology's past, if they became more widely known, would have a significance far beyond our particular concerns at this meeting. Such interactions provide an excellent precedent for the future development of greater overall cooperation among physicians, veterinarians, and biologists[5] for the fuller realization of the enormous potential for medical progress that resides in such partially overlapping, interdisciplinary areas of science as comparative medicine/comparative pathology, pathobiology, medical zoology, and veterinary public health.[6,7]

The main reason this example is such an especially apt one on which to build "one medicine" for the future is that interprofessional cooperation in parasitology, uniquely among biomedical disciplines, has characterized the *clinical* and *preventive* aspects of the field almost as much as its basic science aspects. Surely in no other area of human clinical or preventive medicine have veterinarians and other animal biologists assumed such active and important functions overall in teaching, research, *and practice* as in human parasitology and tropical medicine. Consequently I am most pleased that an interdisciplinary, interprofessional meeting such as this has been organized.

In surveying the past of American parasitology one is struck immediately by how much of the leadership in fostering this unique atmosphere for cross-professional cooperation was taken by just three institutions: First, the Bureau of Animal Industry (BAI) of the United States Department of Agriculture,* and then the Zoological Laboratory (Department of Zoology) of the Land-Grant University of Illinois and the Department of Medical Zoology (and its successors) of the School of Hygiene and Public Health of the Johns Hopkins University. Significant contributing roles were played, too, by other institutions, most notably perhaps McGill, Harvard, Tulane, California, and several land-grant universities; the Rockefeller Institute (now University) and other Rockefeller interests; and the National Institutes of Health. It is with some of the early personalities and germinal events within these several institutions that we shall be concerned here.

Were we to begin by identifying the one individual most directly responsible for launching parasitological *research* in America it would surely have to be Daniel Elmer Salmon, founding director in 1883 of the BAI. An important factor influencing the establishment of the BAI, not only as a pioneering disease control agency, but as America's first institute for experimental medicine, was an embargo placed by several European countries on the importation of American swine and pork products because of the danger of trichinosis. Another was the grave threat posed to our growing beef cattle industry by Texas fever, then a disease of unknown etiology. Of significance to us, therefore, was that Salmon's initial act as the director of BAI was to set up this country's first microbiological research laboratory,† and, immediately thereafter, an

---

* Called the Veterinary Division from 1883 to 1884, the BAI as such was abolished in the course of the governmental reorganization carried out during the Eisenhower administration.

† As early as 1879 Salmon had actually established a more modest microbiological laboratory to support his field studies.

experiment station with holding and handling facilities for small laboratory and large farm animals.

For over a year Salmon, virtually alone, carried on research and launched his successful eradication campaign against contagious bovine pleuropneumonia. Then in 1885 he appointed Theobald Smith as his first laboratory assistant and W. H. Rose as superintendent of the BAI's Experiment Station. Smith, an 1883 graduate of a two-year clinical program at Albany Medical College, had had prior training at Cornell in microscopy and histology under Simon Gage and in veterinary medicine under James Law.* Frederick L. Kilborne, a Cornell agricultural and veterinary graduate and classmate of Smith, soon succeeded Rose at the Experiment Station. By 1885 Salmon's staff, in addition to Smith and Kilborne, totaled sixteen disease investigation and control officers (assistant veterinarians) assignable to the field, two bureau agents in the West, a clerk, a messenger, and a laborer.[9]

Salmon had been trained in bacteriology in France at the dawn of that science, had been a member of the first class to graduate from Cornell University, and its first graduate in veterinary medicine.[10,11] Despite the diverse demands on his time in the early years of the BAI—which make accounts of his career sound like those of several very busy men—Salmon continued to work in the laboratory, teaching Smith the bacteriological culture methods newly developed by French and German workers. Almost overnight the two of them joined the ranks of the illustrious international leaders, Salmon having invented a new culture apparatus with which he and Smith isolated and characterized the type-species of the genus *Salmonella,* and then broke completely new ground by demonstrating the immunizing properties of killed bacteria. With this earliest work on porcine salmonellosis the youthful BAI was catapulted to the scientific fore nationally and abroad.

The BAI's parasitological work began next with the appointment on 1 August 1886 of Cooper Curtice, a young Cornell contemporary of Smith and Kilborne, whose charge was to

---

* When Law's veterinary program at Cornell evolved into the New York State Veterinary College in 1894, Gage became one of its first faculty members.[8]

establish a zoological laboratory. Having taught comparative anatomy after receiving his veterinary degree and having carried out paleontological field research for the United States Geological Survey, Curtice, who roomed with Smith, was assigned by Salmon to study a number of helminth and arthropod parasites in sheep and cattle, which were suspected of being the cause of heavy economic losses to ranchers. By 1887 Curtice had already published papers on fascioliasis in cattle and on hydatid disease; these were soon followed by reports on his important work on sheep tapeworms and myiasis in cattle.

Between 1879, when Salmon had been asked by the United States commissioner of agriculture to investigate livestock diseases in the southern states, and 1883 when he founded the BAI, most of his work had been on Texas fever. Salmon quickly became convinced that its epidemiology could not be explained by any then known mode of disease transmission, that is, its spread could not be by means of direct or indirect contact or by water or food. As he studied the behavior of the disease in the field, Salmon realized that an accurate delineation of the northern limit of Texas fever across the country might well throw light not only on its unknown etiology, but on its mode of transmission.

This was soon proven when, in deciding on the initial efforts of the BAI against Texas fever—and despite young Smith's objections—Salmon approved Kilborne's request for permission to investigate the prevalent theory of stockmen that ticks were responsible for the disease. Salmon's willingness to direct some of his very meager resources to the serious investigation of a layman's theory, which John Gamgee, a leading worker in England, had considered almost too ridiculous for comment, was influenced by his own observation that the northern limits of the disease *and of ticks* were beginning to appear confluent. That decision, among several others equally astute, was to firmly establish Salmon as one of America's leading scientific administrators.

While Smith pursued his work on the pathology and possible microbial etiology of Texas fever, and Kilborne tested the tick theory, Curtice was directed to start work on the biology of the tick itself. By the time the first year of Kilborne's experiments had demonstrated that the cattle tick was essential for the transmission of Texas fever,[12-14,*] Curtice had established that, in contrast to earlier descriptions of the life cycle of a tick, the Texas fever tick spent its entire larval, nymphal, and adult life on the body of a single bovine host.[†]

When Curtice resigned in 1891 from the first of his three periods of service with the BAI, Salmon appointed Charles W. Stiles, a young German-trained zoology Ph.D., to replace him as assistant in the Pathological Division, with responsibility for its Zoological Laboratory. With this appointment Salmon once again displayed his unusual adroitness in selecting very gifted young men for the multiple jobs the BAI was being charged to undertake. This is an important point for us to recognize, for at a time when professional protectionism was still the rule in both human and veterinary medicine—Louis Pasteur, for example, continued to be attacked by the French medical establishment as a "chymiastor . . . poaching on the preserves of others"[15]— Salmon was busily appointing to his staff, as was Pasteur, not only veterinarians, but able physicians, chemists, and zoologists. His early vision of the fruits "one medicine" was capable of producing, together with the wise, encouraging direction and relative freedom of action he gave his subordinates,[‡] created within the BAI an atmosphere highly conducive not only to first-quality research, but to cooperative, multidisciplinary efforts toward practical solutions of major medical problems. In short, the BAI was from its inception an innovative and highly successful medical research institution that quickly attracted international notice to American work in parasitology and related fields.[§]

---

* K. Jacklin, 1976: personal communication from Cornell University, Department of Manuscripts, University Archives, and Fred L. Kilborne papers (1884–1936).

† It was this particular feature of its life cycle that later caused Curtice to advance the completely novel proposal that Texas fever could best be combated through an attack on the tick. He was therefore the father of vector control.

‡ Salmon was unusual for his time in not putting his own name first—or even at all—on most of the reports issuing from his laboratories.

§ My own survey some years ago of American acquisitions in major European human and veterinary medical libraries showed the oldest to be the publications of these earliest BAI scientists on swine salmonellosis—then thought to be hog cholera— and bovine piroplasmosis (Texas fever).

For many years the Zoological Division of the BAI remained by far the largest employer of parasitologists in America. Stiles and his successor Brayton H. Ransom, another zoologist, continued the quality of the original parasitological work initiated by Curtice, and in this they and the division chiefs who followed, such as Maurice C. Hall and Benjamin Schwartz, were associated over the years with a creative and highly productive group of veterinarians and zoologists, including, among others, Albert Hassall, H. W. Graybill, Gerald Dikmans, Allen McIntosh, Eloise Cram, B. G. Citwood, J. T. Lucker, Willard H. Wright, M. W. Horsfall, W. H. Krull, E. W. Price, J. S. Andrews, and Joseph E. Alicata.[16,17] These men and women pioneered in studies to elucidate the life cycles and other ecological complexities of important helminthic parasites. Among the practical systems for prevention and control of parasitic infections that developed from these efforts were those used with success against hookworm, bovine piroplasmosis, equine venereal trypanosomiasis, scabies, cysticercosis, and swine ascariasis.

Other noteworthy examples of the many contributions of the BAI to the subsequent development of American parasitology was the amassing in its laboratories and in the United States National Museum, largely through Hassall's efforts, of one of the largest reference collections on parasites in the world, and the creation by Hassall and Stiles of a key, and still in many respects unique, medical research tool, the *Index-Catalogue of Medical and Veterinary Zoology*.[18] It should also be noted that, as part of its nationwide campaign against equine venereal trypanosomiasis (dourine), BAI scientists developed the first semiautomated serological techniques, which launched seroepidemiology as a discipline, and that the BAI's last director, Bennett T. Simms, like its first, left his mark on American parasitology.*

---

* In the Department of Veterinary Science of Oregon State University, which he had headed, Simms directed the team that studied salmon poisoning in dogs and discovered in the process the most ecologically complex infectious cycle yet disclosed in nature, one in which a parasitic fluke with successive snail, fish, and mammalian hosts serves as the vector for transmission of a rickettsia, which Simms believed originally to be a virus.

From the standpoint of the subsequent development of human parasitology/tropical medicine in the United States, however, the most germinal of the research contributions to parasitology made by the BAI were, next to its investigations on Texas fever, those on hookworm disease. The discovery by Stiles of the hookworm *Necator americanus* and of the wide prevalence of human hookworm disease in the South and the later development by Hall of the first effective means—carbon tetrachloride and then tetrachlorethylene—to combat human hookworm disease on a wide level had profound effects.[19-21]

Of equal importance as these research activities, however, was the indirect influence this veterinary institution was to have on the development of institutions of human medicine. An important manifestation of this was the godfather role the BAI assumed for the nascent National Laboratory of Hygiene, commonly referred to as the Hygienic Laboratory, of the United States Marine Hospital Service, which was moved to Washington from Staten Island in 1891—in 1930 it became the National Institute of Health.

When the advisory board of the Hygienic Laboratory was created by an act of Congress in 1902, in addition to medical experts from the army and navy, its members, not surprisingly, included others from the BAI.[22] That legislation directed that, were qualified physicians not available, other competent persons were to be employed "to take charge of the divisions of chemistry, zoology and pharmacology." This made possible a more direct BAI role in the future of the Hygienic Laboratory and reflected an extension of the refreshing interdisciplinary precedent Salmon had fostered earlier. Stiles, who had just made his important hookworm discoveries, moved from the BAI to head the newly created Division of Zoology of the Hygienic Laboratory.

Stiles and Hassall continued to collaborate on their *Index-Catalogue,* and in 1904 Stiles described a new species of tick, *Dermacentor andersoni,* in connection with his new laboratory's investigations of Rocky Mountain spotted fever. A more important role for the Hygienic Laboratory was realized in 1909, however, when, at Stiles's urging, John D. Rockefeller, Sr. set up the Rockefeller Sanitary Commission with a donation of $1 million for a five-year "eradication program" against hookworm disease in the South. The idea for the Sanitary Commission had grown originally out of Rockefeller's growing interest in the education of Negroes in the South.

Later in this chapter further attention will be given to the hookworm campaign and to its importance to parasitology in America, but first we should note the appearance on the scene of a second, highly creative and farsighted individual who, like Salmon, was indirectly to profoundly influence the future development of human parasitology/tropical medicine in America. I refer to William H. Welch, first dean, in 1893, of the medical school of the newly established Johns Hopkins University, a physician who also had had close veterinary connections and interests. As a young doctor, Welch had journeyed twice to Europe, where he first studied comparative anatomy under Rudolf Leukart in Leipzig and, in 1884, veterinary pathology and microbiology in the laboratories of Theodor Kitt at the Munich veterinary school.[23] Upon his return Welch began to work on swine salmonellosis. Continuing investigations initiated by Salmon and Smith at the BAI, he reported his results to the International Veterinary Congress of 1894. By 1902 Welch was also serving, with BAI scientists, on the first advisory board of the Hygienic Laboratory, the board that had initiated Stiles's appointment in medical zoology.

Before considering Welch's more important influence on the future of human parasitology per se, it should be noted that the Hygienic Laboratory turned again to the BAI when Stiles retired in 1936. Hall, who had held the identical post in the BAI after Ransom's death in 1925, in 1936 became the second chief of the Division of Zoology of what was now the National Institute of Health (NIH).[24-26] Hall had been a pioneer in the development of rational methods to critically evaluate the several substances then being used widely as anthelmintics in animals and man. While still at the BAI he next turned his research interests to possibilities for new chemotherapeutic agents against parasites. Thus Hall must assuredly be considered the father of rational therapeutics in helminthology.

Having been the first to correlate physical and chemical properties of potential drugs with their anthelmintic properties, Hall proceeded from evaluative studies of chloroform, then in common medical/veterinary usage, to discover the efficacy of carbon tetrachloride against hookworms—and also to suggest initially, and establish subsequently, its toxicity to the host's liver. A systematic search for a superior, less toxic drug among related halogenated hydrocarbons had then led to his discovery of the anthelmintic value of tetrachlorethylene. It was this drug that was to provide the means for successfully combating hookworm disease, not only throughout the American South, but elsewhere in the world, and that has since led to the development of many other chemically related and efficacious anthelmintics.

A widely read philosopher of science and a creative scientific leader, Hall was the first of two former BAI veterinarians who were to head the NIH Division of Zoology over the next twenty-three years of its history. The second was one of Hall's first appointees in the NIH, Wright, who had worked with him on anthelmintics in the BAI and had established that the efficacy of different halogenated hydrocarbons was closely related to their water solubilities. Under Wright, the division was redesignated the Division of Tropical Diseases and expanded greatly in numbers of staff and the importance of its program. Wright continued the extensive work on trichinosis initiated during Hall's tenure, and at the advent of World War II the division branched out into important studies on the wide range of parasitisms that would be encountered by American armed forces in foreign theaters of war.

Wright's personal interests then turned largely to schistosomiasis, and it was under his direction and with his active participation that major United States involvement in schistosomiasis research was undertaken. Among the laboratories he established at the NIH were those of parasite immunology and biochemistry. Wright was later a guiding hand in the development of the rapidly expanding activities of the World Health Organization (WHO) in the field of schistosomiasis. Thus for the first fifty-seven years of the existence of federal research activities in human parasitology/tropical medicine in the United States, these efforts were directed by veterinarians and zoologists who had already established their scientific reputations within the veterinary research arm of the government.[27]

We will turn back at this point to the influence on research and training in medical parasitology of the Rockefeller interests and, with them, that of William Welch. In 1911, two years after the Sanitary Commission's initial hookworm grant to Stiles, the enormous sum of $100 million was set aside by another newly created Rockefeller Foundation instrument, the International Health Board, for global work on malaria and on hookworm disease.[28] Not only was the board to be instrumental over the years in training many of the world's malariologists and in supporting pioneering efforts at malaria control, but it and other Rockefeller Foundation sources encouraged and supported parasitological training and research generally, much of it through the creation and financial support of most of the world's schools of public health, in which human parasitology/tropical medicine represented important activities.

In 1901 Welch, along with Smith, had been appointed to the initial board of directors of the planned Rockefeller Institute for Medical Research.[29] Smith had left the BAI in 1895 to head the Massachusetts State Board of Health laboratories, with a concurrent appointment as Harvard's first professor of comparative pathology, one of the first professorial appointments in veterinary medicine by an American school of human medicine. The Rockefeller Institute was planned, largely by Welch, to become the equivalent, under human medical aegis, of the research arm of the BAI and of similar medical research institutes abroad founded by men such as Pasteur, Robert Koch, Jean-Baptiste Auguste Chauveau, and Sir Arnold Theiler.* As part of his campaign to interest American physicians in such an institute for experimental, that is, animal, medicine, and in the fruits of interprofessional cooperation—the

---

* We should note in this connection that other movements in experimental or comparative medicine in America were already well underway. As examples, a veterinary journal concerned in large part with reporting experimental medical research, the *Journal of Comparative Medicine and Surgery,* had commenced publication in 1880. The school from which Curtice had obtained his veterinary degree some time before was called the Columbia Veterinary College and School of Comparative Medicine. Another New York veterinary school had a similar title and program emphasis, and in 1889, at William Osler's urging, the Montreal Veterinary College was redesignated the Faculty of Comparative Medicine and Veterinary Science of McGill University and began to offer the D.C.M. (doctor of comparative medicine) degree.

"one medicine" of Rudolph Virchow—Welch addressed the American Medical Association (AMA) in 1904 on "The Bureau of Animal Industry: Its Service to Medical Science."[30]

Welch's candidate for the new Rockefeller Institute directorship was Smith, who was soon approached to head its bacteriology department, visualized as becoming the core of the developing institute. Smith proposed an organizational structure for the institute that would have included a laboratory for the study of animal parasites, but, as he declined the laboratory position then offered him, a parasitology laboratory was not realized until Smith joined the institute in 1914 as head of its new Department of Animal Pathology located in Princeton.

One of Smith's first recruits to the department was Graybill, a veterinarian and parasitologist from the Zoological Division of the BAI. Following up on Curtice's observation that "blackhead" (histomoniasis) of turkeys occurred only when they were raised in proximity to chickens, Graybill soon made the unprecedented discovery that a nematode, *Heterakis gallinae*, was essential to interhost transmission of the protozoal agent of the disease, *Histomonas meleagridis*, which had been described earlier by Smith. (This protozoan did not produce clinical illness in chickens.) The exact role of the nematode in transmission was later determined by Ernest E. Tyzzer, then in Smith's former department at Harvard. The growing parasitological reputation of the Rockefeller Institute was enhanced considerably through the efforts of Norman R. Stoll, who joined its Department of Animal Pathology in 1927 after obtaining the Sc.D. under William W. Cort at Johns Hopkins.[31]

Welch was also instrumental in the establishment in 1914 of the Hooper Foundation for Medical Research at the University of California and in the appointment of its first two directors, George F. Whipple, a protege of Welch, and, later, the then young veterinarian Karl F. Meyer, who had moved from the veterinary faculty of the University of Pennsylvania to the University of California in 1913 as associate professor of tropical medicine.[32] Meyer, who had worked under Theiler in his famed laboratories in Onderstepoort, quickly established the Hooper Foundation's preeminence in experimental medicine on the West Coast. The indirect but highly important role of

Welch in American parasitology/tropical medicine was now becoming quite apparent.

As founding dean of the Johns Hopkins medical school, Welch had joined with Osler and others to quickly build that school into a leadership position in American medical education. Hopkins was not only the first American medical school to have trained researchers heading all of its preclinical departments, but, under Osler, whose own medical research interests had begun with studies on trichinosis in James Bovell's laboratories at the Ontario Veterinary College, research also became for the first time a part of clinical medicine.[33,34]

Before the innovative changes effected at Hopkins had begun to exert their influence overall, human and veterinary medical education in America had shared different, but equally unfortunate, influences from England. The progressive programs in human medicine eventually instituted at Hopkins, Michigan, and Harvard were therefore closely related to German precedents in medicine. Insofar as human parasitology in America was concerned, the most direct German influence was assuredly exerted by Leukart of Leipzig, the great nineteenth century zoologist. Not only had Welch studied with him, but Stiles had obtained his A.M. and Ph.D. degrees under Leukart. Moreover, Henry Baldwin Ward, the trainer *par excellence* of parasitologists in the early years of the field in America, had also spent time in Leukart's laboratory before receiving his Ph.D. from Harvard in 1892.*

In contrast to the predominant German influence on human medical progress in this country, veterinary medicine benefited more by escaping its initial English legacy from individuals and examples in Scotland and France. Law at Cornell, originally from the Royal Veterinary School in Edinburgh, had been an important shaper of the career of Salmon and had taught Smith, Curtice, and Kilborne. Another more direct effect of the Edinburgh school on North American parasitology

---

* Leukart had published the first edition of his text on human parasitology for physicians and zoologists in 1867. It was translated into English in 1886 and henceforth was to be a major influence on the education of medically oriented zoologists in America. Leukart was a leading and forceful exponent of zoological input into medical and veterinary education, research, and practice.

was strongly manifested through McGill University and its affiliated Montreal Veterinary College, which had been founded by the Scottish veterinary microbiologist, Duncan McEachran, an early mentor of Osler. Osler later held a joint professorship at the college and at the McGill medical school before moving on to Pennsylvania and, later, to Hopkins. Salmon, like several other leading American veterinarians, had studied in France, and the immigrant French veterinarian Alexandre Liautard became another strongly positive influence in American veterinary medicine.

With the exception of a few of the better and long since defunct proprietary schools and the few North American veterinary schools founded in association with schools of human medicine, such as Pennsylvania, McGill, and Harvard, progress in American veterinary medicine occurred under the land-grant university system, that is, it developed in closer association with the goals of agriculture than with those of human medicine. Because parasitology has always been a major subject in veterinary education it was almost inevitable that some of the parasitological departments and laboratories of greatest importance to the future of human parasitology/tropical medicine would be established within these agriculturally oriented universities, which were so uniquely American and practical in their conception. Ward is the one individual whose name stands out most prominently in this connection, particularly for his role in educating future American parasitologists.[35,36,*]

Ward's early career was spent at the Land-Grant University of Nebraska. Appointed in 1893, he eventually headed its Zoological Laboratory, and, uniquely as a zoologist, served also from 1899 to 1909 as dean of its medical school.† A prior development at Nebraska of importance to our theme, but of less direct moment to parasitology, had been the establishment in 1886 by Frank S. Billings, a German-educated American veterinarian, of its Pathobiological Laboratory, which he di-

---

* M. Brichford, 1980: personal communication from University of Illinois, university archivist.

† Two years of scientific training were provided by Ward in Lincoln and two years of clinical medicine in Omaha at a private medical school taken over by the university.

rected until 1893.* As an individual of cross-professional bent, best known for his book, *The Relation of Animal Diseases to the Public Health*,[38] Billings offered some of the earliest instruction in microbiology to American physicians and veterinarians.[39,40]

The list of leaders in parasitology trained by Ward, first at Nebraska, and between 1909 and 1933 when he headed the Zoological Laboratory at the University of Illinois, is a long one. At Nebraska it included the BAI's third and fourth chiefs of parasitology, Ransom and Hall, as well as George R. La Rue, who moved with Ward to Illinois, where Ernest Carroll Faust, James E. Ackert, Myrna Jones, Henry Van Cleave, Horace Stunkard, Thomas Magath, J. E. Guberlet, G. W. Hunter, Harold Manter, R. L. Mayhew, Cort, Justus Mueller, and Paul Beaver were among his and his colleagues' students.

In 1918, again through Welch's efforts, the first American school of public health, another especially important contributor to the advance of parasitology in this country, was established at Hopkins. Parasitology began there with the appointment as chairman of Robert William Hegner, a gifted young zoologist who established the Department of Medical Zoology, which soon became one of the world's most active centers of research and training in parasitology.[41] Forty individuals were to receive their doctorates in parasitology under Hegner's personal direction during the next twenty years, among them a number of future leaders.

Cort joined that department in 1919 as associate in helminthology, and his training role was eventually to become even greater than that of his chairman.† He had spent the two previous years as helminthologist in the Division of Parasitology of the Hygienic Laboratory (later the Public Health Laboratories) of the California State Board of Health, where he worked

---

* The term pathobiology, which now fairly commonly designates an area for cross-disciplinary veterinary/medical/zoological studies, was coined by Billings. There is no evidence that he and Ward ever met (Billings left the year Ward arrived) or that the one laboratory related in any way to the other. (J. G. Svoboda, 1980: Personal communication from University of Nebraska, university archivist.) Unfortunately, Billings was a rather polemical antagonist of Salmon, just as, apparently, no love was lost between Ward and Stiles.[37]

† M. Brichford, 1980: personal communication from University of Illinois, university archivist.

with C. A. Kofoid and encountered William Herms, a creative force in the development of medical entomology, and Meyer. The California Hygienic Laboratory, a research-oriented pioneer among state diagnostic laboratories, had been founded earlier by A. R. Ward, a veterinarian, early food microbiologist, and professor of bacteriology at the University of California. It was in that laboratory that Cort undertook his first studies of hookworm disease, an interest he was to carry with him to Hopkins to such great advantage.

If we were to identify the two lines of parasitological research that most clearly established an international leadership position in parasitology for America and that played the biggest roles in forging the fruitful medical/veterinary/zoology links with which we are concerned here, we would surely select the hemaprotozoal infections, particularly piroplasmosis and malaria, on the one hand, and hookworm disease and related strongyloses on the other. For quite soon after Perroncito's epochal discovery at the Turin veterinary school that epidemic anemia among the St. Gotthard tunnel workers was hookworm disease, and that this parasitism was therefore not just an obscure tropical disease in some of Europe's colonies, came the equally important discoveries of Stiles and the efforts to which we have already referred that they directly and indirectly stimulated.

Probably as important to hookworm control as Hall's discovery of tetrachlorethylene were the results of a series of investigations on many aspects of hookworm's epidemiology and host-parasite relations that were initiated at Hopkins by Cort and his many students, including D. L. Augustine, Stoll, W. A. Riley, R. B. Hill, and F. K. Payne, who later emerged as important leaders in the field.

A basically modest man, Cort, like Salmon, failed to put his own name on much important work toward which he had made generous and often essential contributions. It was as a teacher of future leaders, however, rather than as a researcher per se, that Cort, like his mentor, Ward, really excelled.[42] The development of this strong research-teaching unit at Hopkins under Hegner and Cort, primarily through their studies on bird malaria and on hookworm disease, clearly established the

"Washington-Baltimore corridor"—the BAI/Hopkins axis—in a position of preeminence in *teaching*, as well as in research in, parasitology in America.

The only other institution time permits me to mention in any more detail in this connection is McGill. Parasitology in Canada is said to have commenced with Osler,[43,44] who carried out most of his parasitological research at the Montreal Veterinary College, which was then the most progressive and exciting school of its kind in North America. Upon the very unfortunate closing of that school in 1903, largely because it was so clearly ahead of its time,[45,46] Osler suggested that, at the least, a department of medical zoology should remain within McGill University, a recommendation that was eventually accepted. A suggestion of Osler's that was followed more immediately, however, was the establishment of the Dominion Animal Diseases Laboratory at Hull, a Canadian version of the BAI.

The medical zoology venture at McGill came to fruition in 1932, however, when another Scottish veterinarian, Cameron, was appointed professor of parasitology and director of the university's new Institute of Parasitology. This first organization in the British Empire devoted exclusively to parasitological research soon established an international reputation and has since produced many prominent leaders in the field. Again, Cameron was a man of "one medicine" vision, and his *Diseases of Animals in Relation to Man*[47] was only the second book in the English language—the first was by Billings—to deal with the subject of veterinary medicine and human health.

I realize that Harvard and Tulane are two of the most important among the several institutions I am necessarily slighting in this brief sketch, but, fortunately, others can comment better than I on the past roles of these institutions in the professional interactions we are exploring here. The present Department of Tropical Public Health at Harvard emerged from a melding of Smith's Department of Comparative Pathology, whose parasitologists included Tyzzer as well as Augustine,* and Richard Pearson Strong's Department of Tropical

---

* Augustine had been Cort's first student at Johns Hopkins. Were we to explore the parasitological genealogies of all of us here today we would of necessity conclude not only that the past linkages between institutions were close, but that most of us

Medicine, which had been founded in the medical school in 1913.[48]

The Peking Union Medical College and the Medical School of the American University of Beirut, which both shared in the Rockefellers' largesse, were also prominently associated with the careers of a number of American parasitologists. Faust who, with Beaver, built on Charles F. Craig's efforts in New Orleans, reached Tulane's medical school via Illinois and Peking, with a sojourn also at Hopkins.

To examine briefly just one last facet of our history, the BAI/Hopkins/Illinois triumvirate was involved equally prominently in the creation of our parasitological societies and journals. The first professional organization of parasitologists in America, the Helminthological Society of Washington, was founded in 1911. Its membership was derived chiefly from the BAI, but it also included Stiles, by then at the Hygienic Laboratory, and a few other helminthologists, such as Nathan Cobb, who were in the Bureaus of Entomology or Plant Industry of the United States Department of Agriculture (USDA).

Meeting at members' homes, the proceedings of the "Helm Soc" appeared first in *Science*. From the society's beginnings, however, Ransom and Stiles had discussed with other members the possibility of publishing a parasitology journal, and from these discussions emerged the *Journal of Parasitology* in 1914 under the editorship of Ward at Illinois. A broader area of interdisciplinary activities in medical zoology than that usually encompassed by parasitology was visualized for this new journal, which was to be devoted to the publication not only of research papers on zooparasites, but, generally, on "the relations of animals to disease." Although its primary emphasis became, and has remained, helminthological, the *Journal of Parasitology* was from its conception definitely interdisciplinary; it was equally interprofessional. This interaction is best exemplified by the fact that early in its life, Ransom, a *zoologist* in a *veterinary* institution was the editor of the journal's very useful note series

---

descend from very few parasitological ancestors and that our discipline has been a highly inbred, if not incestuous, one. As Augustine's last student, I not only can trace my own lineage back to Ward, but I had the rare privilege of being the last occupant of Smith's old laboratory at Harvard before it was extensively remodeled.

called "New *Human* Parasites." The first editorial board included individuals associated not only with the parent institutions in Washington and Baltimore, but with Illinois and other land-grant universities, the Army Medical Corps, and the medical schools; its intentionally wide geograpic spread included McGill.

The Helminthological Society of Washington henceforth published its proceedings in this new journal, as did the American Society of Parasitologists (ASP) from its founding, also in Washington, in 1927. The organizing committee for the ASP had consisted of Ransom and Cram from the BAI and Hegner and Cort from Johns Hopkins. Its initial slate of officers, again widely representative, included Ward as president, Samuel Taylor Darling as vice president, Cort as secretary-treasurer, and Kofoid, Riley, Tyzzer, Hegner, Ransom, P. S. Bartsch, and Stiles as council members.

The Helm Soc, which by then was meeting regularly at Johns Hopkins as well as in Washington, began to publish its own separate proceedings in 1934. The initial editorial committee was made up of one representative each from the Bureaus of Entomology and Plant Industry of the USDA and Johns Hopkins and two from the BAI.

So much then for this important and in some respects unique *multiprofessional* aspect of our past. What might it portend for the future? In his chapter in this book Eli Chernin comments on the much-less-realized possibility for genuine cross-disciplinary efforts in the future—efforts in which parasitologists, whatever their professional origins, may interact with biochemists, immunologists, molecular biologists, and others. As I have been one of those who risked flack from classical parasitologists in years past by being so unesthetic as to store my own parasite collection in forms such as acetone powders, I am intensely interested in new prospects for creating fully qualified immunologists, biochemists, and molecular biologists who not only work with but *know something about* animal parasites. For now it seems the shoe is being transferred to the other foot, and those of us who were audacious enough to dabble in such occult realms as physiology—dare I say biochemistry?—in the past are about to be "reclassified" and catch flack from the other camp.

While all of this inevitable—surely desirable (?)—tempest in a test tube is going on, I would urge that in the process we not lose sight of this equally important interprofessional uniqueness of our past. Elsewhere I have urged that persons be trained who *combine* some of the knowledge, outlooks, and skills present M.D., D.V.M., and biomedical Ph.D. degree holders possess, and have suggested that such individuals are sorely needed as researchers throughout almost all aspects of "general medicine." Perhaps it may be from *professional* stems sprouted from such new educational roots as I outlined in some detail in 1978[49] that exceedingly valuable *interdisciplinary* hybrids such as biochemist/parasitologists, immunologist/parasitologists, and molecular biologist/parasitologists might best be budded off.

For reductionism in parasitology as in other disciplines must surely be tempered by origins in a new medical holism if science is not to lose its vital contacts with people and their real needs and become what its detractors are already accusing it of being. In closing, I therefore commend to you W. I. B. Beveridge's recently published sequel to his long classic *The Art of Scientific Investigation.*[50] Entitled *Seeds of Discovery,* it illustrates beautifully how the chance juxtaposition of ideas, observations, and perceptions from the most diverse of sources has provided most of the truly creative sparks of science.[51]

Beveridge correctly points out that most philosophers of science, including Karl Popper, have restricted themselves to only one facet of what scientists actually do, ignoring this really creative aspect completely. T. S. Kuhn[52] and Håken Törnebohm[53] tell us, also, that this is what most scientists are guilty of in another way, in that we generally ignore *creative* questioning of the established paradigms of our particular research fields until a truly crisis situation has already well developed: that is, a prevailing "world-picture" is seen as faulty; a prevailing methodology or strategy appears no longer suited to problems regarded as important; and a previously prevailing ideal or view of one's science is seen as outmoded and to be leading the science astray. It is too bad necessary changes in sciences have had to occur in that episodic manner and that our only occasional self-examining process is not a continuing one.

It seems clear from the selection of some of the people at

this meeting, as well as from previous discussions, that in parasitology/tropical medicine our paradigms of methodology and strategy, and hence of training, are being brought into much-needed question. Good. But surely not to be lost sight of in this process is the wide applicability throughout medicine of the very productive facet of our past I have endeavored to explore in this paper, through which human parasitology/tropical medicine holds out an example for the future to the rest of human medicine, namely, a functioning view of "one medicine" in which participants in almost the entire researcher-practitioner complex are drawn from multiple traditions and are provided the circumstances and opportunities to interact in the public's interest. It seems to me that this is particularly critical to any discipline that has long been, and continues to be, intimately involved with the global social process we call development.[54]

Surely one way to envisage such serendipitous events in the future as the discovery of vector transmission of infections is not only to encourage unfamiliar *disciplines* to mesh their efforts, but to encourage different but meaningfully related professional traditions to rub shoulders and interact. In this respect parasitology *has had* valuable experiences most other medical disciplines have lacked.

NOTES

1. R. Hoeppli, *Parasites and Parasitic Infections in Early Medicine and Science* (Singapore: University of Malaya Press, 1959).

2. C. W. Schwabe, *Veterinary Medicine and Human Health*, 2nd ed. (Baltimore, Maryland: Williams and Wilkins, 1969).

3. C. B. Philip and L. E. Rozeboom, "Medico-Veterinary Entomology: A Generation of Progress," in *History of Entomology*, ed. R. F. Smith, T. E. Mittler, and C. N. Smith (Palo Alto, California: Annual Reviews, 1973): 333–60.

4. H. E. Meleney and W. W. Frye, "Teaching and Research in Parasitology and Tropical Medicine in Medical Schools in the United States," *American Journal of Tropical Medicine and Hygiene* 5 (1955): 769–75.

5. M. C. Hall, "The Wide Field of Veterinary Parasitology," *Journal of Parasitology* 16 (1930): 175–84.

6. Schwabe, *Veterinary Medicine* (see note 2).

7. _____, *Cattle, Priests, and Progress in Medicine*, Fourth Spink Lectures in Comparative Medicine (Minneapolis: University of Minnesota Press, 1978).

8. S. H. Gage, "The History of Veterinary Education in New York State up to 1916," *Journal of the American Veterinary Medical Association* 92 (1938): 809–16.

9. U. G. Houck, *The Bureau of Animal Industry of the United States Department of Agriculture, Its Establishment, Achievements and Current Activities* (Washington: Privately printed by author, 1924).

10. "Daniel E. Salmon, D.V.M., Chief of the Bureau of Animal Industry," *Journal of Comparative Medicine and Surgery* 8 (1887): 257–62. (It is of interest that this biographical sketch of Salmon is followed by an editorial on recent British medical and veterinary studies on scarlatina signed by William Osler.)

11. W. H. Hoskins, "Salmon Memorial Address," *American Veterinary Review* 48 (1916): 456–59.

12. Houck, *Bureau of Animal Industry* (see note 9).

13. M. C. Hall, "Theobald Smith as a Parasitologist," *Journal of Parasitology* 21 (1935): 232–43.

14. G. Dikmans, "In Memoriam, Cooper Curtice, 1856–1939," *Journal of Parasitology* 25 (1939): 516–20.

15. R. Vallery-Radot, *The Life of Pasteur* (New York: Doubleday, Page, 1909).

16. A. O. Foster, "In Memoriam, Benjamin Schwartz, 25 November 1889–21 October 1976," *Journal of Parasitology* 63 (1977): 1110–11.

17. _____, "Centenary. Biographical Note, Brayton Howard Ransom, 1879–1925," *International Journal of Parasitology* 10 (1980): 87–88.

18. E. W. Price and G. Dikmans, "In Memoriam, Albert Hassall, 1862–1942," *Journal of Parasitology* 29 (1943): 232–35.

19. B. G. Schwartz, "A Brief Resume of Dr. Stiles' Contributions to Parasitology," *Journal of Parasitology* 19 (1933): 257–61.

20. C. W. Stiles, "Early History, in Part Esoteric, of the Hookworm (Uncinariasis) Campaign in Our Southern United States," *Journal of Parasitology* 25 (1939): 283–308.

21. W. H. Wright, "Charles Wardell Stiles, 1867–1941," *Journal of Parasitology* 27 (1941): 195–201.

22. R. C. Williams, *The United States Public Health Service, 1798–1950* (Washington: Commissioned Officers Association, 1951).

23. Schwabe, *Progress in Medicine* (see note 7).

24. B. G. Schwartz, "Obituary, Maurice Crowther Hall," *Science* 87 (1938): 451–53.

25. B. G. Schwartz and P. D. Harwood, "Maurice Crowther Hall as a Parasitologist," *Journal of Parasitology* 24 (1938): 283–90.

26. E. A. Chapin and J. S. Wade, "Obituary Notice: Maurice Crowther Hall," *Proceedings of the Entomological Society of Washington* 40 (1938): 147–48.

27. W. H. Wright et al., *Tropical Health, A Report on a Study of Needs and Resources* (Washington: National Academy of Sciences/National Research Council Publ. 996, 1962).

28. Williams, *U.S. Public Health Service* (see note 22).

29. G. W. Corner, *A History of the Rockefeller Institute, 1901–1953, Origins and Growth* (New York: Rockefeller Institute Press, 1964).

30. Schwabe, *Veterinary Medicine* (see note 2).

31. M. S. Ferguson, "Norman Rudolph Stoll: Scientist, Teacher, Friend," *Experimental Parasitology* 41 (1977): 253–71.

32. J. B. de C. M. Saunders and E. B. Shaw, "Karl Friedrich Meyer, 1884–1974," in *In Memoriam* (Berkeley: University of California, 1976): 80–86.

33. D. A. Murphy, "Osler, Now a Veterinarian!," *Canadian Medical Association Journal* 83 (1960): 32–35.

34. C. H. Eby, "Sir William Osler and Veterinary Medicine—A Biographical Sketch," *Journal of Small Animal Practice* 1 (1960): 276.

35. W. W. Cort, "Professor Henry Baldwin Ward and the *Journal of Parasitology*," *Journal of Parasitology* 18 (1932): 99–105.

36. J. R. Christie, "In Memoriam, Henry Baldwin Ward, 1865–1945," *Journal of Parasitology* 32 (1946): 323–24.

37. E. C. Faust, Letter of 29 January 1964 in the Henry Baldwin Ward File, University of Illinois Archives.

38. F. S. Billings, *The Relation of Animal Diseases to the Public Health* (New York: D. Appleton and Co., 1894).

39. Schwabe, *Veterinary Medicine* (see note 2).

40. ————, *Progress in Medicine* (see note 7).

41. W. W. Cort, "Robert William Hegner, 1880–1942," *Journal of Parasitology* 28 (1942): 175–77.

42. Faust, Letter in Ward File (see note 37).

43. T. W. M. Cameron, "The McGill University Institute of Parasitology," *McGill News* 14 (1933): 15–20.

44. ————, "Veterinary Education in Canada," *Veterinary Journal* 93 (1937): 102–06.

45. "The Montreal Veterinary College and Its Founder and Principal," *Journal of Comparative Medicine and Veterinary Archives* 9 (1888): 78–84.

46. Schwabe, *Progress in Medicine* (see note 7).

47. T. W. M. Cameron, *Diseases of Animals in Relation to Man* (London: Faber and Gwyer, Ltd., 1926).

48. J. C. Bequaert, "In Memoriam, Richard Pearson Strong," *Journal of Parasitology* 34 (1948): 515–17.

49. Schwabe, *Progress in Medicine* (see note 7).

50. W. I. B. Beveridge, *The Art of Scientific Investigation* (London: Heinemann, 1957).

51. ————, *Seeds of Discovery* (London: Heinemann, 1980).

52. T. S. Kuhn, *The Structure of Scientific Revolutions* (Chicago: University of Chicago Press, 1970).

53. H. Törnebohm, *Queries About Inquiries* (Göteborg, Sweden: University of Göteborg, Department of the Theory of Science, Rpt. no. 117, 1979).

54. T. Nordenstam and H. Törnebohm, "Research, Ethics and Development," *Zeitschrift allgemein Wissenschaft* 10 (1979): 54–56.

THOUGHTS ON
THE HISTORY OF PARASITOLOGY

*Eli Chernin**

Appropriately enough, we began this conference on the present status and future of parasitology by considering the past. This exercise in history gives us perspective and even lends a certain air of benediction to these proceedings.

History is a peculiarly perishable product. History in the making does not yet exist in comprehensible form; it is akin to the fetus in utero. But history once born begins to perish, for if time lends retrospective insight it also distorts and ossifies. Given enough time, writers, and revisionists, history sometimes bears the same resemblance to reality that a formalin-fixed liver section does to living tissue. But because history represents the path from the past to the future we must try to mark that path for our successors. Knowing history may not prevent mistakes, but ignorance of history invites them.

If we accept the fact that the history of a country is affected by its terrain and by the proclivities of its people, then, in the first instance, the history and development of parasitology rests on the inherent nature of the beasts we study, and, in the second, on the scientists who study them.

What is it about the parasites that governs research and hence the history of the field? The worms are large, complex,

* Recipient, Research Career Award, National Institute of Allergy and Infectious Diseases, United States Public Health Service.

multicellular organisms living in the "warm secret gloom" of the host. Yet more: the helminths multiply deviously, but not in their definitive hosts; their cycles are complex; and they usually refuse to live in laboratory animals or in axenic, defined cultures. While some parasitic protozoa are better behaved, few have yielded to our blandishments, and upon these we now lavish our lucre and our most advanced talents—hence the explosive emphasis on the hemoflagellates and the plasmodia.

If the Swiss Alps and the English Channel have helped fashion history, the inherent characteristics of the animal parasites have fashioned the history of parasitology. Although this is obvious, few nonparasitologists, and not many parasitologists, grasp it. I hasten to add that, despite the constraints imposed by built-in ecological needs, by the reproductive quirks of the parasites, and by the difficulty of maintaining many of them, parasitologists have nonetheless generated a valuable body of scientific information.

Relatively few parasites are available for study with newer methods, and on them rests the history we build today. Focusing on a few parasites will gain us a critical mass of insight, as in obvious examples from genetics and molecular biology, but parasitologists must work with parasite X or Y—not out of choice, but because there is nothing else available. We are not in a position to choose a specific parasite to help answer important questions.

Instead the parasite is there, like the proverbial mountain, to be studied for whatever information it grudgingly yields up. Efforts sharply focused on a few organisms may yield dividends, but a hazard exists that the descriptive base of the field may disappear, leaving sophisticated investigators to search for new targets: Trypanosoma brucei rhodesiense "doth not a summer make," for all its biological beauty, importance, and fecundity.

In the early days of parasitology, publications ran heavily to taxonomy, morphology, life histories, and clinical or related studies; biochemistry, physiology, immunology, and now molecular biology lagged behind, however, partly because of limitations imposed by the number and nature of available organisms. It follows that parasitologists have until recently been primarily preoccupied with parasites rather than processes.

Once scientists venture to inquire into processes, then choosing the organism likeliest to yield an answer becomes a first concern; if the choices are too restricted the experienced investigator may return to *E. coli*, assuming, of course, he considered studying parasites in the first place.

But if the immunologist or molecular biologist is unwilling or unable to make the field grow at its base, by gut-scraping animals killed in the field or people who died in hospital, by learning to tell a trypanosome from a tapeworm, by untangling ascarids from angleworms, or by engaging in fecal alchemy, then others must do this and receive support. Failing this, those chosen few at the so-called "cutting-edge" will not have much to slice from. Many of these matters involve both history and the future, upon which general thought T. S. Eliot wrote:

> Time present and time past
> Are both perhaps present in time future,
> And time future contained in time past.

As I suggested earlier, the history of parasitology has been affected by parasites *and* by those studying them. As the field grew beyond taxonomy and morphology, parasitologists began to apply new techniques. Although a few workers possessed extraordinary talents and special training—Theodor von Brand and Ernest Bueding come to mind—most classically trained parasitologists struggled to establish themselves in unfamiliar technical areas. In some instances, and medical malacology is a case in point, parasitologists eventually dominated the subspeciality because few others cared about trematodes or their snail intermediaries.

Those researchers with a biochemical bent worked alone or occasionally recruited the needed talent, but for years their small tribe was anathema to the Establishment. After World War II, existing journals, still conventional in attitude, disavowed the upstarts, but the advent of new academic programs and new publications broke the fetters. So we have a history of outreach, mostly by conventionally trained parasitologists whose research carried them into uncharted waters where navigating and learning to navigate went on simultaneously. If some publications proved unspectacular one could hardly expect otherwise. Recall that the recruitment of workers specifically trained, say, in biochemistry, was virtually at a standstill, whether for lack of interest or opportunity I cannot say. Clearly, however, parasitologists did not and do not interdict hybridizing their field with others.

So the parasites limited the nature of research, parasitologists were limited in their acquisition of new technical know-how, and few nonparasitologists with the required capabilities were drawn to the field. While the corpus of research involving a plethora of parasites and problems generated a modest literature, it did not generate much theory, the element that has loomed so large in other sciences. Although some disagree with his view, Albert Einstein once said, "It is the theory that decides what you can observe," not the other way around. Without theory, to borrow from Stephen Vincent Benét, "We don't know where we're going, but we're on our way."

Happily, parasitology is now moving rapidly along modern lines and is attracting many scientists from other disciplines. The evidence of change is all around us: new journals devoted to parasite biochemistry, immunology, and molecular biology; more such papers in established journals; and greater availability of funds for research in these subjects, if not in others.

A most interesting development is the course on "The Biology of Parasitism," first taught at the Marine Biological Laboratory in Woods Hole during the summer of 1980. Here a blend of parasitology, immunology, and molecular biology was presented by an awesome battery of invited experts, six days a week for six to ten weeks. Obviously no one learned everything, but three main barriers were breached: the language of the several fields became familiar and hence nonthreatening; handling and maintaining parasites dispelled their mystery or capacity to intimidate; and basic techniques were taught and used. As one perceptive colleague put it, the course not only taught, it "entrapped, excited, and corrupted" many into parasitology. The general theme of the course, I might add, emphasized processes rather than parasites. The course succeeded, by all accounts, and I was glad to have participated in such a novel undertaking.

It is perhaps worth remembering that parasitology's modern history in America is brief. The American Society of Parasitologists dates only from 1924, which just happens to be my own birth year, and I am but the fourth generation of our breed, having descended from Gilbert Otto via William W. Cort and Henry B. Ward. Other colleagues, some still active, count themselves among the third and even the second generation of American parasitologists. Nevertheless, it is a sobering thought that we have no written history of parasitology, and I hope some day to correct that flaw. The field has grown and contributed to the mainstream of biology, and we have worked beyond our discipline, slowly developing new skills or importing them. These developments from inside and from outside are both important, for keep in mind that while immunology grew up mainly *within* microbiology, biochemistry grew by *invading* microbiology.

In any event I would hope, indeed I would insist, that the reductionists now firmly in control at the bench, at the journals, and at the cash register remember that the soups they study come from intact animals, and that holistic biology must have its place alongside the molecular biology of the parasite. This tension between "old" and "new" fields will carry us into the future and countervail stagnation. But replacing one field with the other, in either direction, serves only those whose view of science and the natural world is hopelessly myopic.

THE STATUS OF PARASITOLOGY:
THE CURRENT SCENE

TEACHING PARASITOLOGY:
THE CURRENT SCENE*

Paul P. Weinstein†

It was with some trepidation that I consented to give a talk on as cryptic a subject as the current scene in the teaching of parasitology. Teaching and curriculum design as we all know are very personal activities not easily analyzed by surveys or by the many varieties of teacher/course evaluation schemes that have mushroomed in modern university life.

Teaching by the individual instructor is to a great extent conditioned by her/his perception of the term parasitology. Historically, parasitologists have been intrigued by the intricacies of parasite life cycles and their relationships to infection and disease. Thus until fairly recently teaching of the subject, as evidenced by a perusal of textbooks, understandably has been dominated by the traditional subject areas of morphology, biology and life cycles, pathogenicity and symptomatology, diagnosis and treatment, and epidemiology and control. This was true of most textbooks, even those for introductory courses; the more advanced and specialized ones increased the amount of detail, but usually followed a similar format. I believe it is fair to say that such an approach has predominated more in the United States than in Europe.

* Supported in part by National Institutes of Health Training Grant AI07030.
† The author thanks Dr. Ralph Thorson and the American Society of Parasitologists for permission to use data obtained by its Committee on Education.

Biology as a discipline, however, has undergone a revolutionary change in the past few decades, particularly molecular, cell, and developmental biology, and this has been accompanied at the organismal, population, and ecological levels by the synthesis of new and far-reaching theoretical concepts. In assessing the intellectual health, vigor, and relevance of parasitology teaching to both the field itself and to the general discipline of biology it becomes necessary to view and analyze it against the backdrop of the enormous advances that have occurred in the understanding of biological processes.

It has become evident that at least some parasitologists feel disquietude and discontent. One concern is that they have narrowed their view of the discipline to a dangerous degree and have slighted the concept that parasitism is only one facet of a universal biological phenomenon. As a consequence of this restricted view, principles of parasitism are too often ignored in the teaching of parasitology. This matter was addressed in . detail in a 1958 report by a committee chaired by Clay G. Huff and sponsored by the National Academy of Sciences (NAS)/ National Research Council (NRC).[1] The report, "An Approach toward a Course in the Principles of Parasitism," departed radically from the traditional course format and explored new approaches to the content and organization of undergraduate courses in parasitology. The ideas expressed in the report are as fresh and stimulating today as they were twenty-two years ago, and I commend it to all who are interested in a challenging approach to teaching.

The report was oriented mainly toward courses in parasitology. The committee expressed the thought, however, that there is clearly a trend toward formulating a biology of symbiosis, which would include the phenomenon of parasitism, and that the basic course in the subject may eventually be one in the more broadly encompassing concept of symbiosis. Clark P. Read, who served on the committee, was stimulated by the experience and the teaching concepts expressed to write the textbook *Parasitism and Symbiology*, published in 1970,[2] which embellished and illustrated in depth A. de Bary's original definition of the term symbiosis within a modern biological framework. It is of interest that William Trager's book *Sym-*

biosis, which appeared in the same year, also emphasized the necessity to view parasitism in a broad biological context.[3]

The NAS/NRC report recognized that there are certain intellectual obstacles to the design of courses of the type it advocated, one being that teachers legitimately fear the detachment from facts that endangers any course based primarily on principles. Most recent writers of parasitology texts appear to have recognized this concern and have generally taken a somewhat middle ground. The concepts of symbiosis and parasitological principles are usually discussed at some length; the major portion of these texts, however, returns to a modified traditional approach with some interspersion of newer contributions from biochemistry, electron microscopy, cell and developmental biology, immunology, genetics, and ecology. These aspects are still being treated rather cursorily, but they do serve to introduce students to the exciting interdisciplinary approaches being taken to elucidate and clarify the complexities of parasitism. More exploratory and varied approaches to textbook writing that incorporate the rapid changes and concepts occurring in the field of experimental parasitology are urgently needed by teachers at both the undergraduate and graduate levels.

The American Society of Parasitologists (ASP) has for many years been concerned with both the teaching of the subject and the role of parasitologists as teachers. These matters have been the subject of two surveys in recent years by the Committee on Education of the ASP. In addition, conferences and symposia on these topics have been held at a number of its annual meetings. In 1974 the Committee on Education, chaired by Ralph Thorson, presented a report to the council of the ASP on various aspects of the teaching of parasitology. The report was not published, and I will attempt to summarize briefly a few of the pertinent findings.

Approximately 52 percent of the 1,582 parasitologists contacted responded to a questionnaire sent out by the committee. The institutional affiliations of the 819 who replied are of some interest because they provide information as to where parasitologists are employed and, to a considerable degree, where teaching occurs. As might be anticipated, educational institutions are the principal employers, with government and industry following in that order (Table 1). The affiliations of the 522 parasitologists in educational institutions indicate that the majority (60.7 percent) are concentrated in departments of biology and zoology (Table 2). The next largest numbers are in departments of microbiology, parasitology, and veterinary medicine (22.8 percent). Surprisingly, only 318 (61.2 percent) of 520 parasitologists in educational institutions teach a course in general parasitology. This figure is somewhat disturbing in that general parasitology apparently is not a part of the biology/zoology undergraduate curriculum at a large number of institutions. The reasons for this have not been pursued, but the troublesome specter arises that many departments of biology may consider that a course in parasitology, as traditionally presented, is not of basic significance to the education of young biologists.

The amount of time devoted by various institutions to the teaching of general parasitology varies considerably. The most common length of a course during an academic year is one semester (230 responses), with a one-quarter course next (fifty-one responses), followed by a course consisting of two full semesters (thirty-three responses); last is a course covering two quarters (six responses).

Forty-one percent of the respondents to the questionnaire indicated that they were the only faculty members teaching parasitology at their institutions. A somewhat ominous sign

TABLE 1. EMPLOYERS OF PARASITOLOGISTS, UNITED STATES
1974

Employer	Number Employed[1]	Percent
Universities	522	63.7
Government[2]	115	14.1
Industry	64	7.8
Other[3]	118	14.4
Total	819	100.0

NOTES: 1) Based on a 52 percent return from 1,582 parasitologists. More realistic figures may be obtained by multiplying these figures by two. 2) Number of parasitologists employed: international (5), federal (93), state (15), local (2). 3) Includes graduate students and postdoctoral fellows.

TABLE 2. DEPARTMENTAL AFFILIATIONS OF PARASITOLOGISTS
EMPLOYED IN EDUCATIONAL INSTITUTIONS IN THE UNITED STATES
1974

| Department | Parasitologists | |
	Number	Percent
Anatomy	5	1.0
Animal sciences	2	0.3
Biology	251	48.1
Dentistry	1	0.1
Division of tropical medicine or public health	20	4.0
Entomology	6	1.1
Microbiology	44	8.6
Parasitology	42	8.2
Pathology	11	2.1
Physiology	3	0.4
Poultry science	4	0.6
Primate center	2	0.3
Veterinary medicine	31	6.0
Zoology	66	12.7
Other	34	6.5
Total	522	100.0

regarding the stability of parasitological curricula is suggested by the fact that only 68 percent of the respondents considered there was a commitment on the part of their institutions to the continued teaching of parasitology; the others indicated varying degrees of uncertainty.

Of the 115 parasitologists employed in governmental institutions, 57 (49.5 percent) stated they were involved to some degree in an educational function, most citing the training of technicians and specialized training courses.

The 522 academic institutional respondents listed 881 courses for which they had prime responsibility, 409 (46.4 percent) of these being in parasitology. Because the grouping of these courses into subject areas of parasitology was not given in detail in the 1974 ASP report, I will present the information obtained from a second survey of the membership of the ASP conducted by a subsequent Committee on Education. This survey was concerned specifically with the spectrum of opportunities for graduate education in parasitology. The report of the committee, issued in 1978,* is in essence a listing of institutions and departments in the United States and abroad in which parasitology is taught, the degrees offered, the members of the parasitology faculty, the titles of the course offerings, and the fields of specialization.

Responses to the questionnaire were obtained from 141 educational institutions in forty-five of the United States and Puerto Rico. The data presented here should be considered minimal because, as in most surveys, some institutions did not respond. A summary of the course offerings is presented in Table 3.

A total of 559 courses in fifty-four different subject areas of parasitology is listed, of which ninety-eight appear to be in general or introductory parasitology taught at the undergraduate level. Thus 461 graduate courses are currently being presented nationwide in fifty-three different subject areas of parasitology. Most of these courses are being offered by various science departments (113), with biology and zoology departments predominating (ninety-six); in descending order are veterinary schools (twenty-seven), medical schools (eighteen), and schools of public health (six). Small numbers of miscellaneous other institutions are also involved in the teaching of parasitology, primarily in very specialized aspects.

Only two agricultural colleges responded, but undoubtedly many others teach courses in parasitology, particularly on plant parasitic nematodes. Most of the individuals teaching such courses apparently are not members of the ASP, however, and thus would not have been included in the survey. This is a good illustration of the unfortunate fragmentation of the study of parasitism that has occurred, which was decried in a recent address by S. D. Van Gundy to the Society of Nematologists.[4] To a somewhat lesser degree this may also pertain to protozoologists.

Time is insufficient to undertake a detailed analysis of the subject areas covered by the graduate courses in parasitology

---

* "Opportunities for Graduate Training in Parasitology," Report of the ASP Committee on Education, 1978. Copies may be obtained from the secretary-treasurer American Society of Parasitologists, 1041 New Hampshire Street, P.O. Box 368, Lawrence, Kansas 66044.

TABLE 3. SUMMARY OF COURSE OFFERINGS IN PARASITOLOGY IN 141
EDUCATIONAL INSTITUTIONS IN THE UNITED STATES AND PUERO RICO
1978

Course	Number of Institutions	Course	Number of Institutions
Acarology	6	Nematology/plant nematology	8
Advanced parasitology	19		
Avian parasitology	2	Parasites of lower vertebrates	1
Biology of larval helminths	1	Parasites of the tropics	1
Biology of parasitoids	1	Parasitic radiology and pathology	1
Cestodology	1		
Chemotherapy and control of animal parasites	2	Pathobiology of invertebrates	1
Clinical parasitology	3	Pathology of parasitic diseases	8
Clinical tropical medicine	2		
Diagnostic parasitology	7	Physiology/biochemistry of parasites	20
Ecology of arthropods of medical importance	1	Principles of parasitism	5
Ecology of parasitism	8	Protozoology	45
Epidemiology and control of parasitic diseases	1	Seminar in parasitology (unspecified)	21
Epidemiology of arthropod-borne diseases	1	Symbiosis	11
Experimental parasitology, including problems in parasitology	25	Techniques in parasitology	9
		Topics in parasitology/tropical medicine	20
Field studies in parasitology	1	Trematodology	1
Fine structure of parasites	4	Tutorials and specialized courses in:	
Fish parasitology	9		
General parasitology (undergraduate)	98	Malaria	5
Helminthology	46	Trypanosomiasis	1
Immunology of parasites, including host-parasite relationships	26	Leishmaniasis	1
		Filariasis	1
		Schistosomiasis	4
Intracellular parasitism	1	Vector ecology	1
		Vector genetics	1
Livestock parasitology	1	Vectors and parasites	1
Marine parasitology	5	Veterinary helminthology	6
Medical malacology	2	Veterinary parasitology, including "advanced"	17
Medical parasitology	35	Veterinary protozoology	5
Medical/veterinary entomology	42	Wildlife parasitology/parasitic diseases	12
		Zoonoses	2
		Total	559

presented at various institutions. The great variety of subject matter as represented by the titles, however, attests to the richness of current teaching as well as to the gaps. Of interest is the offering of courses in parasite biochemistry and physiology, parasite ultrastructure, parasite immunology, and symbiosis. Relatively few such courses would have appeared on a comparable list compiled ten years ago.

With regard to federal agencies, the Fish and Wildlife Service and the Centers for Disease Control offer course work/training in parasitology—the former in fish parasitology and the latter in diagnostic procedures.

Various biological field stations and laboratories offer courses in parasitology, usually during the summer months. These courses provide important training, particularly in the areas of life cycle studies, ecological parasitology, and systematics. Of greater significance, they involve students, often for the first time, with parasitism as a natural phenomenon rather than as a laboratory artifice. Descriptions of these laboratories and their course offerings have appeared in recent issues of the ASP *Newsletter*.

Although there are many other pertinent aspects to the teaching of parasitology, I wish to make one particular point. If parasitology is to remain as a vital and important part of the biological scene we must attract our share of the minds and hearts of the brightest of the young biologists when they are undergraduate and beginning graduate students. It is at this time that the choice of professional fields is being made, and as a consequence their impressions and perceptions of parasitology as a challenging discipline assume considerable importance. These young people are being stimulated and intellectually excited in their varied course work and readings by the revolution that has occurred in recent years in the understanding of biological processes and in the application of extremely sophisticated methodologies that have been developed for their study. They encounter this in all their coursework, be it in molecular, cell, and developmental biology, immunology, genetics and evolution, ecology, or population biology. They must therefore also find these elements deeply incorporated in their study of parasitism, so that they see this pervasive biological phenomenon interpreted in terms of contemporary biology rather than only at the usual, more descriptive level. If we fail in this aspect of our teaching, the study of parasitology will be in danger of becoming a backwater.

Lest I be misinterpreted, I in no way mean to imply that parasitology be taught only at the molecular and cellular levels. Parasitism is one of the most complex phenomenon to comprehend, and it is the crossroads of innumerable biological disciplines—this is what makes it so fascinating.

For the past 100 years we have accumulated a rich heritage of information from the beautifully detailed studies of the life cycles and developmental patterns of a myriad of parasites. The elucidation of parasite morphology and life cycles will continue to be an important area of study for many years to come as there is still a great deal to be learned, not only of biological interest, but of particular significance to public health. For example, it is only in recent years that the mystery of the coccidia-*Toxoplasma-Sarcocystis* group of organisms has begun to be resolved, and cycles of such pathogens as the *Angiostrongylus* group, *Capillaria philippinesis* and *Naegleria fowleri*, have been comprehended. Teaching in such classical areas as parasite life cycles, systematics, and evolution must continue to be a basic cornerstone to the discipline of parasitology. Of the parasitological morphologists, systematists, and evolutionists being trained today, however, how many are also being taught to use the tools of molecular and population genetics, DNA and isoenzyme biochemistry, statistics, and computer applications that have proved to be of immense value to systematists in other fields?

It is imperative that in our teaching and in our texts we offer our students much more than a descriptive analysis at the morphological level, otherwise we will fail them seriously. The succession of stages in the life cycle of African trypanosomes, for example, is understood only superficially without a detailed comprehension of the complex changes in surface membrane composition and antigenicity that are manifested as the parasite goes through its succession of morphological types in the mammalian and insect hosts. The accompanying cytological changes that take place in the mitochondrion and kinetoplast

have little more than descriptive significance without an understanding of the concomitant profound biochemical shifts occurring in energy metabolism. In teaching, the stress on life
cycles at the descriptive level must be integrated with an analysis of the interplay of the many physical and physiological
"triggers" of the host and environment that induce membrane
changes and biochemical events in parasites that are ultimately
translated into the gross morphological transformations we
recognize as the development of different morphological types
or successions of polymorphic larval stages.

This approach is not a rejection or denigration of our
parasitological heritage, but merely an acceptance of the fact
that for well over a decade the study of parasitism has been
entering a new era. There is no value judgment involved here.
Rather it is the recognition that new tools, concepts, and theories are now available with which to analyze developmental
processes of parasites, and the interaction of host and parasite,
to provide insights into the complexities of parasitism hitherto
undreamed of. In fact parasites have become objects of research by increasing numbers of nonparasitologists who have
seized upon them as magnificent models to investigate a wide
array of biological processes of fundamental significance to the
fields of biology and medicine. Such a phenomenon is not new
in the history of parasitology. We have only to recall that some
of the most seminal landmarks in biology were made through
study of the horse ascarid, *Parascaris equorum;* namely, the discovery and significance of meiosis, the phenomenon of deter-
minate cleavage and cell lineage, and chromatin diminution. I
fully expect that modern parasitology will also generate comparable biological landmarks.

What is the import of all this for the teaching of parasitology? I believe it signifies that one can no longer be complacent
about teaching life cycles and systematics in a purely descriptive manner. It is necessary to interpret these phenomena
within the context of contemporary biology. I suggest that
parasitology students should be thoroughly educated in such
subjects as molecular biology and biochemistry, cell and developmental biology, endocrinology, immunology, and neurobiology and neurochemistry. The importance of training in the

understanding of the organization and function of surface and internal cell membranes cannot be overemphasized. Obviously parasitologists will not be able to become expert in all these fields, but they should have sufficient exposure to as many of them as possible so they can later apply the information and concepts gained to their research and teaching.

Parasite life cycles do not occur in a void, however, and the phenomenon of parasitism is not expressed only at the molecular and cellular levels. Parasitology to a great degree is an ecological phenomenon, and the peregrinations of parasites outside the host are as complex to analyze as those within the host. The parasitologists of today studying ecological aspects of parasitism require thorough training in such areas as ecosystem and community structure, population biology and population genetics, coevolutionary theory, mathematics, statistics, and computer application. A solid background in such disciplines should permit ecological parasitologists to undertake the quantitative analytical studies required to begin to understand parasite population dynamics and to formulate the theoretical base required for sound epidemiological studies.

The teaching of parasitology in the years ahead will undoubtedly vary considerably as different departments experiment and develop approaches compatible with their interests. Obviously no single format will be universally acceptable, nor would that be even remotely desirable. Students should find varied academic options in parasitology open to them, ranging in emphasis from the molecular to the population level. Because such parasitological curricula by their very nature are basically interdisciplinary, however, certain problems tend to arise. The quality of the teaching and research of parasitologists working in the area of molecular biology, for example, will inevitably be judged by other biologists on the basis of the denominators that establish competence for the general field of molecular biology and by the standards of the best journals in that field. The same will also hold true for the various other combinations of disciplines, such as immunology and ecology, with parasitology.

These quality judgments of parasitologists by their peers are of professional and practical importance in that they relate directly to the way in which parasitologists are perceived by the biological community and to the way in which young parasitologists as applicants for positions are viewed by department chairmen and appointments committees. It follows inexorably from this that parasitology must stand not only as a strong discipline in its own right, but must provide its graduate and postgraduate students with a thorough background in the various fields of contemporary biology so they may compete successfully and make significant contributions in interdisciplinary areas. I view this as the primary challenge for the teaching and training of young parasitologists today.

NOTES

1. C. G. Huff, L. O. Nolf, R. J. Porter, et al., "An Approach toward a Course in the Principles of Parasitism," *Journal of Parasitology* 44 (1958): 28–51.

2. C. P. Read, *Parasitism and Symbiology* (New York: Ronald Press, 1970).

3. W. Trager, *Symbiosis* (New York: Van Nostrand Reinhold, 1970).

4. S. D. Van Gundy, "Nematology—Status and Prospects: Let's Take Off Our Blinders and Broaden Our Horizons," *Journal of Nematology* 12 (1980): 159–63.

TEACHING VETERINARY PARASITOLOGY IN AMERICAN UNIVERSITIES

Michael J. Burridge

SOURCES OF DATA

In preparation for this conference I sent a short questionnaire to the thirty-nine colleges and schools of veterinary medicine and departments of veterinary science in the United States (Table 1). The purpose was to gain basic information on the teaching of veterinary parasitology at American universities.

Three basic questions were asked:

• What courses do you offer in veterinary parasitology?

• In what years of the professional veterinary curriculum are the parasitology courses taught?

• How many student-contact hours are there in each parasitology course, and how are those hours divided between lectures and laboratory work?

All but three of the institutions responded to one of the two questions; those not responding were Cornell University, South Dakota State University, and the University of Missouri.

A more detailed survey, "Trends in Veterinary Parasitology Instructional Programs in North American Veterinary Medical Curriculum: 1957–1976," was prepared for the American Association of Veterinary Parasitologists (AAVP) by Helen E. Jordan. The president of the AAVP, Edward L. Roberson, kindly sent me the general information section and the summary section of the report of that survey.

TABLE 1. COLLEGES, SCHOOLS, AND DEPARTMENTS
OF VETERINARY MEDICINE IN THE UNITED STATES
1980

University of Arizona, Department of Veterinary Science, College of Agriculture
Auburn University (Alabama), School of Veterinary Medicine
University of California (Berkeley), School of Veterinary Medicine
Colorado State University, College of Veterinary Medicine and Biomedical Sciences
Cornell University, New York State College of Veterinary Medicine
University of Florida, College of Veterinary Medicine
University of Georgia, College of Veterinary Medicine
University of Idaho, Faculty of Veterinary Medicine
University of Illinois, College of Veterinary Medicine
Iowa State University, College of Veterinary Medicine
Kansas State University, College of Veterinary Medicine
University of Kentucky, Department of Veterinary Science, College of Agriculture
Louisiana State University, School of Veterinary Medicine
University of Maine, Department of Animal and Veterinary Sciences
University of Maryland, Department of Veterinary Science, College of Agriculture
Michigan State University, College of Veterinary Medicine
University of Minnesota, College of Veterinary Medicine
Mississippi State University, College of Veterinary Medicine
University of Missouri, College of Veterinary Medicine
Montana State University, Department of Veterinary Science, College of Agriculture
University of Nebraska, Department of Veterinary Science, Institute of Agriculture
    and Natural Resources
University of Nevada, School of Veterinary Medicine, College of Agriculture
North Dakota State University, Veterinary Science Department
Ohio State University, College of Veterinary Medicine
Oklahoma State University, College of Veterinary Medicine
Oregon State University, School of Veterinary Medicine
Pennsylvania State University, Department of Veterinary Science, College of Agricul-
    ture
University of Pennsylvania, School of Veterinary Medicine
Purdue University, School of Veterinary Medicine
South Dakota State University, Department of Veterinary Science
University of Tennessee, College of Veterinary Medicine
Texas A & M University, College of Veterinary Medicine
Tufts University, School of Veterinary Medicine
Tuskegee Institute, School of Veterinary Medicine
Utah State University, Department of Animal, Dairy, and Veterinary Sciences, College
    of Agriculture
Washington State University, College of Veterinary Medicine
West Virginia University, Division of Animal and Veterinary Sciences, College of
    Agriculture and Forestry
University of Wisconsin, Department of Veterinary Science, College of Agricultural
    and Life Sciences
University of Wyoming, Division of Microbiology and Veterinary Medicine, College of
    Agriculture and Agricultural Experiment Station

RESULTS OF THE SURVEY

Courses in Professional Veterinary Curricula

A summary of preclinical parasitology courses taught to veterinary students in colleges and schools of veterinary medicine in the United States is given in Table 2. The courses range in number from one to three and are taught between the first and third years of a four-year curriculum. In general the courses cover basic veterinary parasitology, followed by a detailed consideration of the major parasitic diseases of animals. The number of student-contact hours in preclinical parasitology varies widely, from a low of fifty-three to a high of 177, with an average of 110 hours. Student-contact hours devoted to

TABLE 2. SUMMARY OF PRECLINICAL PARASITOLOGY
COURSES TAUGHT TO VETERINARY STUDENTS IN
AMERICAN COLLEGES AND SCHOOLS OF VETERINARY MEDICINE

Institution[1]	Year(s) of Veterinary Curriculum in Which Course(s) Taught	Student-Contact Hours		
		Lectures	Laboratory	Total
Auburn	2	70	40	110
California	1 & 3	50	60	110
Colorado State	1–3	83	30	113
Florida	2–3	34	30	64
Georgia	2	65	70	135
Illinois	2	60	90	150
Iowa State	2	70	40	110
Kansas State	2	45	75	120
Louisiana State	2–3	61	67	128
Michigan State	2–3	39	48	87
Minnesota	2	60	60	120
Mississippi State	1–3	90	36	126
Ohio State	1–3	49	4	53
Oklahoma State	2	75	60	135
Pennsylvania	1	46	39	85
Purdue	2	83	71	154
Tennessee	1–2	50	25	75
Texas A & M	1–2	87	90	177
Tufts	1–3	42	31	73
Tuskegee	2	30	60	90
Washington State[2]	2	60	45	105

NOTES: 1) Neither Cornell nor Missouri responded. 2) Students at Idaho and Oregon State take parasitology courses at Washington State.

laboratory sessions also vary considerably—from 8 percent to 63 percent.

It is impossible to compare with any accuracy the teaching of clinical parasitology in veterinary colleges, schools, and departments due to the integrated nature of clinical instruction in the veterinary curriculum.

Graduate Courses

Graduate courses in veterinary parasitology offered by American institutions are listed in Table 3. Seminars and problem or research-credit courses are not included.

Five veterinary colleges of schools—California, Idaho, Mississippi State, Tennessee, and Tufts—and seven veterinary

TABLE 3. GRADUATE COURSES IN VETERINARY PARASITOLOGY
OFFERED BY AMERICAN VETERINARY COLLEGES, SCHOOLS, AND DEPARTMENTS

Institution	Course Title	Credit Hours
Arizona	Parasites of Domestic Animals	
Auburn	Helminthology I	5
	Helminthology II	5
	Protozoology	5
Colorado State	Medical Helminthology and Protozoology	
	Veterinary Parasitology	
	Parasitic Disease Control	
	Foreign Diseases of Domestic Animals II (Parasitic Diseases)	
	Parasitic Diseases of Wildlife	
Florida	Veterinary Parasitology I	5
	Veterinary Parasitology II	5
	Helminthology	5
	Immunology of Animal Parasites	5
	Parasitic Diseases in the Tropics and Subtropics	5
Georgia	Pathology of Parasitic Diseases of Animals	2–5
	Helminthology	5
	Veterinary Nematology	5
	Techniques in Experimental Parasitology	5
	Immunity to Animal Parasites	
Illinois	Experimental Parasitology	
Iowa State	Pathogenic Protozoa	5
	Pathology of Parasitic Diseases	5
Kansas State	Advanced Veterinary Parasitology	3

Institution	Course Title	Credit Hours
Louisiana State	Veterinary Helminthology	4
	Veterinary Protozoology	3
	Chemotherapy and Control of Animal Parasites	3
Maryland	Parasitic Diseases of Domestic Animals	
Michigan State	Parasitic Zoonoses	
	Parasitic Diseases of Wildlife	
Minnesota	Parasites of Wildlife	3
	Advanced Veterinary Parasitology	
Montana State	General Parasitology	
	Topics in Animal Parasitology	
Nebraska	Veterinary Parasitology 903	
	Veterinary Parasitology 904	
Ohio State	Immunology of Parasitic Infections	4
	Applied Veterinary Parasitology	2
Oklahoma State	Techniques in Parasitology	
	Advanced Helminthology	3
	Biology of Parasites	
	Parasitic Protozoa	3
Oregon State	Veterinary Parasitology	
Pennsylvania	General Parasitology	
	Immunology of Host-Parasite Relationships	
	Parasitic Zoonoses (Protozoa)	
	Parasitic Zoonoses (Helminths)	
	Biology of Parasites (Adaptation to Parasitism)	
	Biology of Parasites (Natural Immunity and Host Specificity)	
Purdue	Pathology of Parasitic Diseases	3
Texas A & M	Parasitology	4
	Host-Helminth Relationships	3
	Parasitic Protozoa	4
Tuskegee	Veterinary Helminthology	3-6
	Veterinary Entomology	3
	Veterinary Protozoology	3
Washington State	Veterinary Parasitology	3
Wisconsin	Helminths of Livestock	4
	Protozoology	4
	Prevention and Control of Parasites of Domestic Animals	3
Wyoming	Veterinary Parasitology	
	General Parasitology	
	Helminthology	

departments—Kentucky, Maine, Nevada, North Dakota State, Pennsylvania State, Utah State, and West Virginia—offer no graduate courses in veterinary parasitology.

DISCUSSION

SHER: Dr. Weinstein, has the current expansion in research in parasitology—the involvement of other disciplines in the field—filtered down to the level of the university? Are students now entering the field aware of the parasitological research going on in the larger institutions? And is this a source of stimulation to them?

WEINSTEIN: Those questions can be answered only if we know how parasitology is being taught, and that is difficult to evaluate on a national level. If you examine most of the textbooks on parasitology, which is probably the more objective way, I believe little of the excitement of current parasitological research has yet penetrated them. This is true for most texts prepared for undergraduate courses. If young, bright students at the point where they are making decisions about their careers are offered only traditional textbooks on parasitology, they may well be turned off. This is an area where the greatest revision in approach has to be made if we are to excite students to enter the field and give them an understanding of the tremendous changes occurring in experimental parasitology. This approach has to be reflected in textbooks and, more specifically, in the way the subject is presented by teachers. We will see a decline in interest in parasitology unless we bring to the attention of students the revolution that is indeed occurring.

PINO: Is anyone in the process of putting together educational material along these newer lines?

WEINSTEIN: As far back as 1958 Clay Huff and his study group made a plea to depart from the traditional course material. Our greatest problem is to get people to write textbooks that reflect what is going on today, not several decades ago.

SHERMAN: I subscribe to Weinstein's point of view. When I started to teach parasitology I was very much influenced by the Huff Report, but in practical terms it was difficult to use such a format in a course, except at an advanced level.

If you try to incorporate the dynamic aspects of parasitism in any undergraduate course it becomes a very expensive process. You have to maintain life cycles and have experience in handling a variety of parasites; in addition, it is very time-consuming. There is a finite amount of time available for students to do some of these experiments, and curricular patterns do not permit them to do so.

Another aspect is that it is sad but true that many parasitologists are not at the cutting edge of the field. By and large, the kind of course presented to students, which is not revealed in surveys, is probably a canned course that provides minimum exposure. That kind of teaching probably pervades the field. The exceptional teaching of parasitology is just that—exceptional. The field therefore takes on the character of the majority; it slides backwards and has a dampening effect.

One reason for the dearth of textbooks is the lag between the time one starts to write a book and the time it is available to students; the current texts were probably generated between five and ten years ago, so many exciting developments we are now aware of do not appear in them. Another factor is that those of us who are interested in developing the field at the research bench don't write textbooks. It is one of those mutually exclusive areas. What we really need is an amalgamation of bench scientists and individuals who can write well.

PINO: Or provide opportunities for scientists to break away for a year or two.

WARREN: I am sure you are aware of Foster's *History of Parasitology*, which was published in England. What interests me about that book is not what is in it, but what is not. It is very descriptive, but the key factor in the whole process of investigation is left out.

Foster clearly shows that since ancient times people have been aware of the visible helminths, and in many cases associated parasites with disease. He also points out that the first so-called parasitic protozoan—*Entamoeda histolytica*—wasn't discovered until 1870. What he never even considers is what brought helminthology and protozoology together in a field called parasitology, or how the discipline evolved and separated from other aspects of parasitism. This is just as interesting as the converse—how bacteriology and virology escaped from protozoology and helminthology.

A new edition of *The Immunology of Parasitism* is in the process of being assembled by Sydney Cohen. I must in all candor say that the current edition does not indicate the great potential of immunology and parasitism. I hope the new edition will attract the attention and interest of bright young people.

KLEBANOFF: While we are talking about how to encourage students to enter parasitology and how to train them adequately for their life's work, I would like to address this issue from the vantage point of the head of a Division of Infectious Disease in a Department of Medicine, where we see people acquire an interest in parasitology rather late.

These individuals often took inadequate courses in parasitology in medical school; had two or three years of house staff training, where they saw little parasitology; and then took an infectious disease fellowship. At this point, perhaps because of a local association, they fell in love with parasitology and wished to pursue it vigorously.

What do we do with these people? Do we send them for a crash course at the London School of Hygiene and Tropical Medicine? Or do we present them with a parasitological problem in the laboratory and expect and hope that, because of their intelligence and concern, they will gradually accumulate knowledge about life cycles, appropriate therapy, and so on? Can we expect them ultimately to have the breadth of a professional parasitologist who has trained continuously in this area from day one? There is a vast pool of such people, who are basically physicians interested in curing disease, and this is a resource that should be tapped. But how do we best train them at this stage?

PINO: You are raising two points: the preparation of professionals to become parasitologists, and the training of people who need to be made aware of parasitology.

KLEBANOFF: No, these people will practice parasitology in their own way, but not in the sense of having an encyclopedic knowledge and overall expertise in the field. Rather, they will dedicate much of their lives to parasitology, having come to it from a broad interest in human or veterinary medicine, not as parasitologists in the classical sense.

DELAPPE: It is difficult to schedule such things. A classic example of this type of evolution is Byron H. Waksman, who entered the field as an immunologist and a physician and who has been making a fine contribution to it. Jack S. Remington is another. He was the first to find that *Toxoplasma* infection evokes a host response of interferon. How can we program the growth of those who develop a later rather than an early interest in parasitology?

CHERNIN: Dr. Klebanoff, when you say people are interested in a further career, is that as medical parasitologists or as physicians? What would they be doing, typically?

KLEBANOFF: Their interests may begin with the organisms we would classify as parasites. Then, by virtue of this concern, they may become interested in the disease they produce and find an outlet for their clinical training either in this country, through the ever-increasing incidence of parasitic diseases we are seeing, or through involvement in developing countries.

CHERNIN: The last route is in many ways the one most of them would like to follow—a career overseas working with parasitic infections or something similar. The problem is that no existing mechanism for that amounts to much; there is no ladder one can climb.

MAHMOUD: To get back to Klebanoff's point, there are two levels of recruitment to parasitology: one is at the undergraduate college level; the other is in medical school. I think he was addressing the latter level. The physicians see parasitology as a discipline of internal medicine called infectious diseases, which encompasses parasitic infection. But for some reason infectious diseases in America have been reduced to viral and bacterial infections.

Some teachers of courses in infectious diseases, such as Klebanoff, are now saying there are other agents that can and do infect man in every part of the world, and that this is an exciting area for research and for training clinicians.

But those who conduct such training programs face a real problem. How can they give students some indication that they are on an equal level with medical graduates in virology and bacteriology? None of these courses can help in that way because we cannot send the students for a month every two years to the London School of Hygiene and Tropical Medicine or to Woods Hole for a six-week summer course.

There has to be some attempt at the medical school level, or later on, to give these people a starting point of interest in parasitology and build that up so that immunology and biochemistry follow.

BOWERS: We went through a period, Dr. Klebanoff, when a large number of elective programs were included in the medical curriculum. A great many students spent a period of time in developing countries working in areas where people had parasitic diseases. Did that have any influence on the career interests of students of parasitology?

KLEBANOFF: I'm sure it did.

CHERNIN: The best training for such people would probably be obtained overseas rather than in this country. I can't imagine what training they would get here.

PINO: From my knowledge of overseas opportunities it would be erroneous to train students specifically for that purpose. We need people who are broad-gauged, with a wide range of interests.

CHERNIN: I have no argument with Burridge and Weinstein about the need to stimulate young people who will be coming into the field, preferably at the undergraduate level. The fact is, however, that it is hard to make parasitology a challenging discipline.

In our discussions, incidentally, we have not mentioned the

financial aspects. Where is the money coming from to train these graduates? Where are the jobs going to be in the future? What career opportunities will there be in X years from now? No matter how exciting we make parasitology sound at the junior or senior levels, if there are no job prospects it becomes an academic exercise.

SIMPSON: Teaching a workshop or conducting courses such as the one at Woods Hole are proper ways to attract people from other disciplines. In the summer of 1980 I participated in a very successful two-week laboratory course on "The Molecular and Cell Biology of Trypanosomes" at the International Institute of Pathology in Brussels. It was conducted by Fred Opperdoes and Piet Borst and had about twenty-one participants from Europe, South America, and Africa.

RESEARCH IN PARASITOLOGY
IN AN EDUCATIONAL INSTITUTION

Julius P. Kreier

The primary function of educational institutions is education; that of government-sponsored and private commercial and industrial research laboratories is the solution of scientific or technical problems.

For any investigative group the problems chosen for study are defined by various mechanisms: political processes that determine the distribution of funds for specific types of research; a board of directors that must decide on what is important in terms of the company's profits; or the concerns of an individual who establishes a trust to create a philanthropic foundation.

The allocation of funds affects the decisions of individual researchers. This may be tempered by the variety of programs from which they may choose and the flexibility of the administration, but the line of investigation is shaped directly or indirectly by the decisions of those who control the funds. Research in an educational institution in the United States is supported by federal, state, or city sources. Public or private educational institutions may be egalitarian or elitist, oriented strongly toward research or not, although an elitist orientation is more likely in a private than in a public institution.

The classification of an educational institution by academic mission appears to correlate more closely with its research activities than does the public-private type of classification. Schools

74

whose primary mission is undergraduate or professional education generally place less emphasis on research than those committed to graduate education. The nature of any institution, be it a school of arts and sciences or of medicine, veterinary medicine, or other professional or technical field, roughly determines the kinds of research developed within it. The degree to which an institution may carry on research appropriate to its educational mission, however, is always strongly influenced by the priorities set by funding agencies.

The research conducted in an educational institution is no less valuable than that done in a research institute; thus one can justify support of research in an educational institution for the same reasons that may be used to justify any scientific research. Educational institutions are designed to educate students in addition to carrying on research; research institutes, on the other hand, designed to produce research results, are not well equipped to educate, except at the postdoctoral level. The dual role of research and training in an educational institution justifies support of its research endeavors and affects the nature of the investigations.

Graduate training is essentially an apprenticeship, and students can be trained to do research only in an environment where it is being conducted.

Few if any educational institutions provide their faculty members with sufficient funds to do the research that is an integral part of their graduate training programs. Thus graduate faculty must compete with other individuals who are seeking funds for research. Attempts to reconcile the requirements of education with research productivity can cause problems for the teaching faculty member who must justify expenditure on education of funds that were awarded for research.

The more advanced the students chosen for research training the fewer the problems encountered by faculty in maintaining productivity in the laboratory. Postdoctoral students are ready to start research promptly on joining the group, and thereafter all their time is so engaged. Graduate students, on the other hand, spend a substantial part of their time in activities connected with their degree programs, which do not directly advance the work in the laboratories in which they are being trained. Most prominent among the activities essential to graduate training, but not directly productive of research, are course work and general and qualifying examinations.

The nature of the educational process affects the research program of the laboratory, just as the research affects the nature of the students' training. A laboratory committed to education is in some respects less flexible than one whose primary function is research. The laboratory head draws up a general plan of research, which is usually incorporated in the proposals submitted to funding agencies, and if approved he may then work with individual students to develop specific parts of the overall project. The rate at which students develop, however, is a major factor in determining when results will be produced. As students generally expect to incorporate their laboratory work in their theses, it is not easy to assign additional people to the problem if the research is not progressing as rapidly as desired. If the director, postdoctoral fellow, or technical assistant should step in and resolve the problem too quickly the students' education and the director's basis for judging their ability will be impaired. It is almost certainly true that people learn best if they have the freedom to make mistakes.

Because research results often form part of a graduate student's thesis, output tends to be irregular. There is, to use the jargon of economics, frequently a long period of capital input with little visible output, followed by a period of intense production with a burst of results. The fact that students complete the program and graduate just as they become valuable to the research project interrupts research productivity. When the trained student graduates, the whole process of training must start over again with a new, untrained individual.

Some years ago I read an article by economists who analyzed the distribution and production of significant contributions to science and technology among laboratories; the data showed that a relatively small proportion of funded laboratories produced most of the tangible results. This supports what most of us have concluded on the basis of subjective analysis. The authors concluded it would be cost effective to concentrate support in the laboratories that produce "results." This conclusion reveals an appalling lack of perception of the way science and education interact, as well as a lack of understanding of psychology and the dynamics of the political process in our society.

Productive laboratories are very much a part of the entire scientific structure. In assembling their staffs they lean heavily on postdoctoral fellows, research associates, and untenured junior faculty who spend a limited number of years doing research with the group. The "less productive" laboratories provide training grounds and are the source from which the leaders of the productive laboratories may draw; they also provide an escape for those who do not wish to stay in the productive laboratories. Thus the latter would soon collapse if the system on which they depend for trained people were to be seriously weakened by lack of support.

If we place too many restrictions on the number of individuals with a stake in science it will be politically impossible to support members of a small scientific elite, even if it were feasible to develop a way to recruit them in the absence of broadly based scientific involvement on the part of our educational institutions.

Duplication and repetition play a greater role in scientific investigation than is often acknowledged. It may appear to be cost efficient and reasonable to limit the number of laboratories working on a given problem, but, in fact, observations are considered accurate by consensus of those working on a given problem. This implies the need for a mass of informed individuals. In science this mass results from the involvement of many people at many levels in the investigative process.

So far this paper has been a highly subjective discussion of the role of educational institutions in research. I will now present some data I have collected on parasitological research in schools of veterinary medicine.

RESEARCH IN VETERINARY SCHOOLS

A questionnaire was sent to twenty-nine veterinary schools in the United States; responses were received from seventeen, or 59 percent, of them. The subset of seventeen is self-selected and represents those that completed and returned the questionnaire; it may therefore not be an unbiased sample. Fifty-nine faculty were identified as parasitologists in the seventeen schools; the average number was three and a half per school; the range was one to eight. Only one school reported having a parasitology department. Most parasitologists are located in microbiology departments, which at times include the disciplines of pathology and public health in addition to parasitology—the term parasitology is not always included in the departmental title. In some schools parasitologists are found in departments of pathology, pathobiology, preventive medicine, and laboratory medicine, or simply scattered among other departments (Table 1).

Most veterinary schools are small, which necessitates faculty of various disciplines being grouped together under one department heading. Departments with two to four faculty, as is the case in many schools, are really too small to conduct graduate training programs and provide modern scholarship. My personal belief is that parasitologists are most appropriately grouped with microbiologists, who in the broad sense are also interested in parasites.

TABLE 1. LOCATION OF PARASITOLOGY FACULTY, BY DEPARTMENT, IN AMERICAN SCHOOLS OF VETERINARY MEDICINE
1980

Department	Number of Faculty	Percent of Faculty
Preventive Medicine	3	18
Microbiology and Parasitology	2	12
Pathobiology	2	12
Pathology	2	12
Laboratory Medicine	1	6
Microbiology	1	6
Microbiology and Pathology	1	6
Microbiology, Pathology, and Parasitology	1	6
Microbiology, Pathology, and Public Health	1	6
Parasitology	1	6
Various	2	12
Total parasitology faculty	17	

NOTE: Based on responses from seventeen of the twenty-nine schools of veterinary medicine in the United States.

The largest single group of responding veterinary parasitologists (46 percent) consider themselves helminthologists; general parasitologists (29 percent) and protozoologists (22 percent) are the next most common types; few entomologists (3 percent) are found in veterinary schools (Table 2).

The large proportion of helminthologists probably reflects the common belief that worms are the most prevalent veterinary parasitological problem in the United States. The teaching requirements of veterinary schools with only one or two parasitologists probably assure the availability of many general parasitologists. Most universities with veterinary schools also have colleges of agriculture, which usually have departments of entomology; thus the veterinary schools may not feel the need for duplication of effort in that discipline.

In addition to being grouped by parasite studied, parasitologists may be classified by techniques used in their work (Table 3). Immunoparasitology, like immunology, is much in vogue today, as is evident in veterinary parasitology: the largest single group of responding parasitologists (32 percent) state that they are immunoparasitologists. Those who consider themselves generalists are also strongly represented (29 percent); these are individuals who adapt techniques as required for their research. Classical parasitologists concerned with morphology and taxonomy are still with us (17 percent). Other disciplines of interest to veterinary parasitologists are epizootiology (12 percent) and biochemical parasitology (10 percent), including parasite metabolism and chemotherapy.

TABLE 2. FIELDS OF INTEREST TO VETERINARY PARASITOLOGISTS IN THE UNITED STATES,
BY TYPE OF PARASITE
1980

	Number	Percent
Helminthology	27	46
General	17	29
Protozoology	13	22
Entomology	2	3
Total	59	100

NOTE: Based on responses from seventeen of the twenty-nine schools of veterinary medicine in the United States

TABLE 3. FIELDS OF INTEREST TO VETERINARY PARASITOLOGISTS IN THE UNITED STATES,
BY TECHNIQUES USED
1980

	Number	Percent
Immunoparasitology	19	32
General	17	29
Classical parasitology (taxonomy, morphology)	10	17
Epizootiology	7	12
Biochemical parasitology (including parasite metabolism and chemotherapy)	6	10
Total	59	100

NOTE: Based on responses from seventeen of the twenty-nine schools of veterinary medicine in the United States.

Very few students undertake graduate training in veterinary schools (Table 4): on the average only one veterinarian is working toward a Ph.D. in each responding school; two schools have a veterinarian working toward a master's degree; while only one in ten reports a veterinarian in postdoctoral training. Graduate students with bachelors' degrees are more common in the responding schools than are veterinarians.

The competition of private practice and positions in industry or government almost certainly draws veterinarians away from graduate work. (My guess is that graduate training does not increase a veterinarian's future earnings—but that would warrant a study in its own right.) I do not believe veterinarians should necessarily undertake their graduate study in veterinary

TABLE 4. STUDENTS IN GRADUATE TRAINING IN PARASITOLOGY
IN AMERICAN SCHOOLS OF VETERINARY MEDICINE
1980–81

Degree	Veterinarians		Nonveterinarians	
	Number	Average[1]	Number	Average[1]
Ph.D.	13	0.9	25	1.8
M.Sc.	7	0.5	16	1
Postdoctoral	2	0.1	4	0.3
Total	22		45	

NOTE: 1) Based on fourteen responses from twenty-nine schools of veterinary medicine in the United States.

schools. A case may be made for the suggestion that training in a medical school or a school of arts and sciences would broaden the veterinarian's experience and, over a period of time, help bring veterinary research into closer contact with other areas of biomedical research. But as the veterinary profession is small it must guard against a loss of identity in its associations with larger and more powerful groups.

Veterinary parasitologists apparently receive support from most of the agencies that fund academic research (Table 5). Among the responding schools, the National Institutes of Health and the United States Department of Agriculture are the largest sources of funds; the United States Army and the World Health Organization also provide substantial support. Funds from experiment stations and state governments are received by almost all schools, but the amounts are small. The data on which Table 5 is based, however, are probably not exact. The questionnaire requested amounts of funds per year per grant, and asked that grants to regular veterinary faculty be

TABLE 5. RESEARCH FUNDING OF PARASITOLOGY IN AMERICAN SCHOOLS OF VETERINARY MEDICINE, BY SOURCE AND AMOUNT 1980

Source	Number of Schools Funded	Amount	Average Amount
National Institutes of Health	6	$1,300,000	$217,000
Department of Agriculture	6	1,086,000	181,000
Department of the Army	2	496,000	248,000
World Health Organization	5	471,000	94,000
Department of the Interior	1	370,000	370,000
Industry	7	360,000	51,000
Experiment stations and state funds	14	235,400	17,000
Agency for International Development	1	102,000	102,000
National Science Foundation	1	40,000	40,000
Other and unidentified	2	119,000	60,000
Total		$4,579,400	

NOTE: Based on data provided by fifteen of twenty-nine schools of veterinary medicine in the United States.

listed separately from grants to associated individuals, such as, for example, those at experiment stations who hold joint appointments in the schools. In many cases it was difficult to determine whether the amounts reported were per year or per grant, and whether they were held by regular faculty or by faculty with joint appointments whose primary responsibilities were elsewhere.

CONCLUSIONS

In conclusion I will attempt to relate the results of the survey of veterinary schools to some thoughts expressed in the first part of this paper.

Veterinary schools in the United States are, with the exception of that of the University of Pennsylvania, parts of state universities. Their traditional academic mission has been professional education, not graduate training. In the past the number of faculty has been hardly more than sufficient to teach the professional students. This orientation toward professional education and the small faculties have inhibited the development of graduate training and research programs. There still seem to be too few parasitologists for optimal development of graduate programs in parasitology in veterinary schools. The few trainees reported by the responding schools seem to reflect the constraints imposed by this shortage of faculty in parasitology.

I was rather surprised at the amount of money reported for parasitology research in the responding schools; the size of the graduate programs seems small in relation to the funding. Even in an educational institution, research can be carried out by a faculty member with the aid of a technical staff and thus be independent of education; this may be the course chosen by those holding grants for research in veterinary parasitology. If that is the case it seems to be a decision that fails to develop to the fullest the potential of veterinary schools as educational institutions. Using a technical staff to do research does, however, spare the faculty member the problem of reconciling the goals of education and research productivity, which was discussed in the first part of this paper.

Another factor discussed earlier was the effect of resource allotment on research faculty and institutions. Strictly on the basis of titles, I scored the grants listed by the respondents according to whether they were connected primarily with veterinary research or with medical research. Of the 106 grants thus rated, the chief focus of seventy-two appeared to be veterinary medicine and that of thirty-four to be human medicine. While I am sure that one-third of grants to medical institutions are not primarily of concern to veterinary medicine, this result indicated more support for veterinary research than I anticipated.

In this brief paper I have presented some of my views on the role of research in an educational institution. I consider that a strong research program is required in any institution attempting to provide graduate training in a scientific discipline, but some very real difficulties face any faculty member who undertakes to reconcile the requirements of education with those of grant-supported research. There are, however, benefits and rewards to the individual who combines education and research. In an educational institution a faculty researcher has the pleasure of working with students who are entering the field with a fresh view, and may get the satisfaction of seeing them become capable scientists with the passage of time.

RESEARCH IN PARASITOLOGY:
THE PERSPECTIVE OF THE
NATIONAL INSTITUTES OF HEALTH

Irving P. Delappe

Just 100 years ago Charles Louis Alphonse Laveran, a French army surgeon then stationed in Algeria, discovered the asexual portion of the cycle of the malaria parasite, in recognition of which he was awarded the Nobel Prize in physiology or medicine in 1907. So it is fitting that we celebrate the anniversary of this discovery by holding a conference of this nature.

While I welcome the opportunity to talk about research in parasitology at the National Institutes of Health (NIH), at the risk of appearing insular and provincial, the title of this paper should be "Research in Parasitology: The Perspective of the National Institute of Allergy and Infectious Diseases" (NIAID).

In the spirit of the research conducted in the early 1900s by Maurice C. Hall on trichinosis and Charles Wardell Stiles on hookworm, the NIAID has been dominant in its support of research in parasitology, medical entomology, and tropical medicine, not only intramurally at the NIH, but extramurally elsewhere in the United States, and to some extent in other parts of the world.

Currently the institute receives over 95 percent of applications addressed to the NIH for support of research and training in these fields. The grant applications undergo initial scientific review in the Tropical Medical and Parasitology Study

Section and are then presented to the National Advisory Council of the NIAID for the policy stage of the dual review process.

Tables 1 through 4 demonstrate the support given by the NIAID for research in parasitology in fiscal years 1978, 1979, and 1980, as well as the various mechanisms of support. Table 3

TABLE 1. NATIONAL INSTITUTE OF ALLERGY AND INFECTIOUS DISEASES
TROPICAL MEDICINE PROGRAM
FISCAL YEAR 1978

Mechanism		Number of Projects	Amount
GRANTS			
Parasitology and Medical Entomology			
Research		148	$ 9,578,770
Training		9	562,393
Fellowships		13	167,816
Careers		8	272,798
		178	10,581,777
Leprosy		6	577,025
Cholera		9	830,948
Arbovirology		19	1,900,309
	Subtotal	212	13,890,059
International Center for Medical Research		4	2,117,000
	Total grants	216	16,007,059
CONTRACTS			
Schistosomiasis		1	118,080
Filariasis		3	136,454
Leprosy		3	230,485
Cholera		3	258,693
Arbovirology		1	51,724
	Total contracts	11	795,436
INTRAMURAL RESEARCH			
Laboratory of Parasitic Diseases (59 personnel)			3,158,000
Rocky Mountain Laboratory (Medical zoology and vector studies; 13 personnel)			402,000
	Total intramural		3,560,000
	Grand total		$20,362,495

TABLE 2. NATIONAL INSTITUTE OF ALLERGY AND INFECTIOUS DISEASES
TROPICAL MEDICINE PROGRAM
FISCAL YEAR 1979

Mechanism		Number of Projects	Amount
GRANTS			
Parasitology and Medical Entolmology			
Research		178	$11,987,080
Training		9	556,628
Fellowships		20	285,200
Careers		7	231,512
		214	13,060,420
Leprosy		9	768,643
Cholera		8	832,559
Arbovirology (includes 2 "tropical virology" for $366,977)		24	2,493,345
	Subtotal	255	17,154,967
International Center for Medical Research		4	2,359,920
International Collaboration in Infectious Diseases Research Program		4	2,035,678
	Total grants	263	21,550,565
CONTRACTS			
Schistosomiasis		2	206,000
Filariasis		2	143,948
Leprosy		6	415,600
Cholera		4	241,524
Arbovirology		1	310,344
	Total contracts	15	1,317,416
INTRAMURAL RESEARCH			
Laboratory of Parasitic Diseases (57 personnel)			2,880,000
Rocky Mountain Laboratory (Medical zoology and vector studies; 12 personnel)			408,000
	Total intramural		3,288,000
	Grand total		$26,155,981

TABLE 3. NATIONAL INSTITUTE OF ALLERGY AND INFECTIOUS DISEASES
TROPICAL MEDICINE PROGRAM
FISCAL YEAR 1980

Mechanism		Number of Projects	Amount
GRANTS			
Parasitology and Medical Entomology			
Research		185	$12,950,000
Training		9	630,000
Fellowships		20	430,000
Careers		7	240,000
		221	14,250,000
Leprosy		10	856,000
Cholera		8	858,000
Arbovirology		27	2,780,000
	Subtotal	266	18,744,000
International Collaboration in Infectious Diseases Research Program		4	2,039,000
	Total grants	270	20,783,000
CONTRACTS			
Schistosomiasis[1]		2	273,176
Filariasis[2]		2	219,662
Leprosy		5	439,559
Cholera		3	239,296
Arbovirology		1	403,851
	Total contracts	13	1,575,544
INTRAMURAL RESEARCH			
Laboratory of Parasitic Diseases (57 personnel)			3,082,000
Rocky Mountain Laboratory (Medical zoology and vector studies; 12 personnel)			437,000
	Total intramural		3,519,000
	Grand total		$25,877,544

NOTES: 1) Office of Naval Research: Genetic characterization of susceptibility and infectivity of schistosomes and snail vectors; Cornell University: Radioactive labeling of schistosomes; University of Lowell, Massachusetts: Maintenance and supply of schistosome-infected snails and mammals—all three human species and vectors. 2) Cornell University: Radioactive labeling of filariae; University of Georgia: Filariasis Repository Research Service—five genera and vectors.

TABLE 4. NATIONAL INSTITUTE OF ALLERGY AND INFECTIOUS DISEASES
INTERNATIONAL COLLABORATION IN INFECTIOUS DISEASES RESEARCH PROGRAM

Principal Investigator	American Institution	Collaborating Country	Host Institution	Title of Grant Proposal
Thomas H. Weller	Harvard School of Public Health	Brazil	Federal University of Bahia	Endemic Schistosomiasis and Chagas' Disease in Brazil
Kenrad E. Nelson	University of Illinois	Thailand	Chiang Mai University	Immunobiology and Epidemiology of Leprosy
Thomas C. Jones	Cornell University Medical College	Brazil	Federal University of Bahia	Basic and Applied Research on Protozoan Diseases
Jeffrey F. Williams	Michigan State University	Sudan	Central Laboratory of Khartoum	Collaborative Research on Parasitic Diseases in Sudan
Thomas C. Orihel	School of Medicine Tulane University	Colombia	Colciencias Bogotá	Selected Vector-Borne Diseases in Man
TROPICAL DISEASE RESEARCH UNITS				
John R. David	Robert Breck Brigham Hospital and Harvard Medical School			Immunology of Fever Tropical Diseases in Man
Adel A. F. Mahmoud	School of Medicine Case Western Reserve University			Immunological Responses to Schistosomiasis and Other Parasitic Infections

indicates that in 1980 awards totaling $14,250,000 were made in support of 221 grants. Most of this money is for free-ranging basic research projects, but it also supports two programs with special emphases:
• The Biological Regulation of Vectors Program
• The Immunology of Parasitic Infections Program
These programs, together with support mechanisms, are listed in Table 5.

The goal of the Biological Regulation of Vectors Program is the advancement of fundamental studies leading to effective methods of vector control other than insecticides and molluscicides. Research conducted under the program involves insect predators, pathogens, and competitors for living space. Examples of current research being conducted in this special emphasis program are:

• The nematode *Neoplectana carpocapsae* holds considerable promise for the biological control of black flies. It is highly virulent to *Simulium*, the vector of onchocerciasis, causing rapid death and high mortality rates. Concern about nontarget susceptibility appears to be largely unjustified. Moreover the likelihood of stream establishment seems remote, given the limited reproductive capacity of the nematode in aquatic invertebrates.[1,*]

• G. Craig reported that seventy female mosquitoes of the large nonvector species, *Toxorhynchites rutilis*, were released in a ten-acre wooded area in Indiana for the control of the vector mosquito *Aedes triseriatus*. They laid their eggs in 136 tree holes in the area and two additional generations of this predator were produced in a three-month period, reducing the *Aedes* population by one-half.[†] The same investigator reported on another accomplishment whereby *Aedes* infected with a new strain of the gregarine protozoan, *Ascocystis barretti*, was released in the field. The mosquitoes were marked with a dye, and ten days later it was demonstrated that the infected mosquitoes had virtually disappeared, whereas uninfected mosquitoes were abundant.[‡]

* R. Gaugler, 1980: personal communication.
† G. Craig, 1980: personal communication.
‡ A conference on the Biological Regulation of Vectors was sponsored by the NIAID in 1975. Copies of the proceedings may be obtained free of charge by writing to the author.

TABLE 5. NATIONAL INSTITUTE OF ALLERGY AND INFECTIOUS DISEASES SPECIAL EMPHASIS PROGRAMS ACTIVE AS OF JUNE 1980, IDENTIFIED BY MECHANISM

| | Research Grant Program | | Training Grant Programs | | | | | | Total Programs | |
| | | | Training | | Fellowships | | Careers | | | |
	No.	Amount	No.	Amount	No.	Amount	No.	Amount	No.	Amount
Sexually transmitted (venereal) diseases	29	$ 3,043,507	1	$ 55,701	1	$ 16,096	2	$ 74,354	33	$ 3,189,658
Biological regulation of vectors[1]	25	1,518,988	2	259,353	2	27,200	1	27,471	30	1,833,012
Persistent infections	18	1,488,459	—	—	—	—	2	74,293	20	1,562,752
Clinical virology	28	1,864,911	3	188,854	—	—	5	148,543	36	2,202,308
Recombinant DNA	41	3,373,614	1	81,403	2	26,400	2	71,437	46	3,552,854
Resistance to antimicrobial agents	39	2,801,316	2	205,541	4	59,498	6	183,176	51	3,249,531
Hospital-associated infections	16	1,413,027	3	337,208	1	17,000	—	—	20	1,767,235
Immunology of parasitic infections[1]	33	3,121,826	5	261,539	4	53,289	1	40,253	43	3,476,907
Antiviral agents, including phage and interferon	32	2,287,795	2	134,040	3	43,180	2	74,672	39	2,539,687
Mycology	40	2,929,512	4	177,748	—	—	2	71,618	46	3,178,878
Streptococcal disease and sequelae	18	1,786,487	—	—	4	43,420	3	113,118	25	1,943,025
Subtotals	319	25,629,442	23	1,701,387	21	286,083	26	878,935	389	28,495,847
Other programs	1,113	101,365,746	61	4,271,497	155	2,034,258	78	2,786,724	1,407	110,458,225
Total all programs	1,432	$126,995,188	84	$5,972,884	176	$2,320,341	104	$3,665,659	1,796	$138,954,072

NOTE: 1) These two special emphasis programs are described in detail in the text.

Major goals of the Immunology of Parasitic Infections Program are studies for the purpose of developing effective vaccines for the prevention of such parasitic diseases as malaria, schistosomiasis, and filariasis; the intervention in the host response to prevent or ameliorate disease processes that are immunologically mediated; and the development or improvement of immunodiagnostic procedures for parasitic infections, especially as they relate to the immune status of the host. Examples of current research in this program are:

• From a study of the relationship of the antifilarial action of diethylcarbamazine and the immune system, evidence indicates that the immune system may be involved in two ways: 1) the diethylcarbamazine may strip a protective coat of immune origin from the microfilariae, thus 2) making it possible for the immune elimination of microfilariae to take place.*

• In a study of the immunological control of schistosome granuloma formation in T cell-deficient mice, serum gamma globulin shows the double effect of 1) protecting recipient liver cells from microvesicular fatty degeneration, and 2) restoring eosinophilic granulocytes to their liver granulomas. Gamma globulin was not absolutely effective in this regard, but whole chronic infectious serum was.†

• Congenitally athymic mice were shown to be significantly more susceptible to *Trypanosoma cruzi* infection than their thymus-bearing littermates, as measured by increased parasitemia, mortality rate, and shortened survival time. In addition, transplantation of neonatal thymus into athymic mice reestablished normal levels of resistance to *T. cruzi*. These results indicate that host defense mechanisms active in experimental Chagas' disease are under thymic control.[2],‡

---

* D. Weiner, 1980: personal communication.

† F. von Lichtenberg, 1980: personal communication.

‡ A workshop on the Immunology of Parasitic Infections was sponsored by the NIAID in June 1977. The proceedings were published in November 1977 as a supplement to the *American Journal of Tropical Medicine and Hygiene*. Another NIAID-sponsored workshop, this time on the application of recombinant DNA technology to protozoology, was held in April 1980. The proceedings were published as a supplement to the September 1980 issue of the same journal. The latter workshop brought together leading molecular microbiologists practicing the occult art of recombinant DNA cloning and protozoologists investigating antigenic variation in African trypanosomes.

Not as an afterthought, but in conclusion, it would most certainly be remiss of me not to mention some examples of the outstanding research being conducted in the Laboratory of Parasitic Diseases of the NIAID.

• Members of the Malaria Section are engaged in an attempt to improve culture conditions for the production of large numbers of *Plasmodium falciparum* gametocytes and gametes. They are also studying the physiology of exflagellation, fertilization, and invasion of red cells by malaria merozoites. Characterization of structure, function, and immunogenicity of surface determinants on gametes and merozoites is being investigated, together with an evaluation of gamete vaccines in model systems for the identification of the best antigens and adjuvants.

• R. W. Gwadz, L. H. Miller, and L. C. Koontz are attempting to develop a simple, rapid, and inexpensive method of screening for antimalarial drugs with activity against hepatic schizonts. They are trying to determine whether drugs that have gametocytocidal activity in a *Plasmodium gallinaceum*/ chicken/*Aedes egypti* system are also active against the hepatic exoerythrocytic stages that cause relapses of certain human and simian malarias.*

• E. A. Ottesen and F. A. Neva, with their collaborators in India, are studying tropical eosinophilia (TE), a form of filariasis characterized by lung disease with episodes of paroxysmal nocturnal asthma, profound blood eosinophilia, markedly elevated IgE levels, and high filarial antibody titers.[3] Using the in vitro correlate of IgE-mediated allergic response—that is, the histamine-release reaction of IgE-coated basophils challenged in vitro with the sensitizing allergen—and a radioenzymatic assay for the detection of histamine, they have documented and quantified the high degree of allergic sensitization to filaria in patients with TE, compared to those with other forms of filarial disease. The special hypersensitization to microfilariae and their products may be most important in determining the symptoms and pathology associated with this syndrome. At the other extreme, individuals with circulating

---

* R. W. Gwadz, 1980: personal communication.

microfilariae, who might also be expected to be allergically sensitized, are paradoxically hyporesponsive to antigens derived from microfilariae. The mechanisms underlying this hyporesponsiveness appear to involve serum inhibitory factors whose nature is under study.*

NOTES

1. D. Molloy, R. Gaugler, and H. Jamnback, "The Pathogenicity of *Neoaplectana carpocapsae* to Black Fly Larvae," *Journal of Invertebrate Pathology* 36 (1980): 302–06.
2. E. Kierszenbaum and M. Pienkowski, "Thymus-Dependent Control of Host Defense Mechanisms Against *Trypanosoma cruzi* Infection," *Infection and Immunity* 24 (1979): 117–20.
3. E. A. Ottesen, F. A. Neva, R. S. Paranjape, et al., "Specific Allergic Sensitization to Filarial Antigens in Tropical Eosinophilia Syndrome," *Lancet* 1 (1979): 1158–61.

* E. A. Otteson, 1980: personal communication.

DISCUSSANT:

Ruth S. Nussenzweig

Irving Delappe's presentation very clearly indicates the nature and extent of the support given parasitological research by the National Institutes of Health, and particularly by the National Institute of Allergy and Infectious Diseases, which unquestionably reviews and funds a cross-section of the very considerable proportion of ongoing parasitological investigations in the United States. It also supports most of the training programs, except those for researchers in developing nations, which are funded primarily by the World Health Organization.

Other more goal-oriented governmental agencies give invaluable aid to certain kinds of parasitological research they consider of primary importance, and/or specific problem areas that are more likely to be solved on a short-term basis by concerted, intensive efforts.

The Rockefeller Foundation, a cosponsor of this conference, has chosen to finance work on a wide range of subjects performed by investigators trained in fields outside the framework of departments of or established groups in parasitology.

Such support is undoubtedly encouraging what has already become a trend during the past few years, namely, the interdisciplinary approach. One can hardly dispute the potential and the appropriate timing of this approach, which indeed reflects the changing realities of parasitological research. It is to be hoped that the application of new methodologies and the innovative reformulation of long-standing problems will advance the state of the art and upgrade the level of investigations in the discipline.

The challenge, I believe, is to find a way to integrate interdisciplinary collaboration with the activities in good parasitological laboratories that are moving in the same direction, but by different routes. This may be achieved by recruiting researchers from other fields and by diversifying research methodologies and directions.

The seriousness of the problem of finding support for these laboratories should not be underestimated as they are responsible for a considerable part of the recent scientific advances in the field. Also, because a substantial number of future parasitologists are being trained at these sites, it is necessary to continue interdisciplinary training in order to assure the production of future researchers. Furthermore, opportunities must be provided to train experts in tropical medicine, otherwise such expertise will disappear rapidly as members of the older generation discontinue their laboratory investigations.

In his presentation Julius Kreier analyzed the status of the parasitologist in an educational institution and demonstrated the interrelationship of research and training activities and some of the difficulties now being faced.

A number of these difficulties have arisen because parasitology is often not an essential component of the curriculum of many professional schools, and therefore does not receive sufficient institutional support: it is frequently tolerated rather than fostered by the universities. In many instances parasitology departments and/or divisions are in fact threatened with extinction.

Let me now present some of my thoughts regarding the present state of parasitology and the directions in which we are moving.

It is an indisputable fact that parasitological research has undergone a radical change in the last ten to twelve years. This is basically due to a shift in emphasis from a more zoological descriptive view to a molecular one that focuses primarily on host-parasite relationships. Emphasis is being placed on investigations aimed at the construction of mathematical models to express biological relationships that clarify the molecular basis of host-parasite relationships. The genetic control of certain parasite traits, as well as mechanisms of host resistance and susceptibility to infection, is proving to be a fruitful new direction.

The biochemical study of parasites has also been given a different emphasis in recent years. Particular attention is now being paid to characterizing the peculiarities of the metabolic pathways of parasites and the mode of action of antiparasitic drugs for the purpose of developing a rational approach to chemotherapy.

The changed focus of parasitological investigations is clearly documented in the present formulation of long-term research goals and in the nature of the problems most frequently investigated.

The rapid rate of significant advances made in the discipline over the last few years has undoubtedly been the result of the application of innovative experimental designs, the reformulation of old problems, and the use of a whole spectrum of techniques available to modern biology. Recent progress is certainly contributing to the present high level of expectations for the early solution of the problems of treatment and control of the most serious and prevalent parasitic diseases.

These diseases continue to take a very high toll in the morbidity and mortality of populations distributed over extensive regions of the globe, and contribute significantly to the maintenance of extreme levels of poverty and underdevelopment in those same regions. Can we as investigators live up to what is expected of us and try to bring closer to reality the conquest of diseases? The answer obviously lies in attempts to develop a basis for effective methods of chemotherapy, vector control, and vaccination procedures.

Should investigators in fact adopt a formulation of research priorities in parasitology or be guided exclusively by their intellectual curiosity and fascination with experimental systems?

I do not consider the two views to be mutually exclusive. Moreover I am certain that part of the attraction of newcomers to the field of parasitology is linked to the potential relevance of their research to the health of people.

I believe the basis for the choice of a research problem in this and other areas is essentially an individual matter, however, and that the choice and its results should be judged primarily on their scientific value. For this reason one can only hope the funding agencies will support and preserve the freedom of choice of individual investigators—a freedom experience has shown to be important for the advancement of science.

One outstanding feature of parasitological research in recent years is the considerable degree to which immunological and biochemical methods, as well as those of genetics and cell biology, are being applied to probe the host-parasite relationship. The benefits of such an interdisciplinary approach are multiple, and very talented individuals are being attracted to such work from other areas of biological research.

This is in sharp contrast to the situation prevailing not too long ago, when an investigator worked in deep isolation. Parasitology was considered to be outside the mainstream of the biological sciences and received no benefit from methodological and conceptual advances in related fields. Parasitology also suffered from geographic isolation, more acutely so at the end of the so-called "colonial era," when it was no longer considered the responsibility of developed nations and their leading scientists to attempt to deal with the most serious health problems of the former colonies.

The link with indigenous investigators in endemic areas is essential, however, from the point of view of research activities and training; the acquisition of in-depth knowledge of the clinical epidemiological characteristics of a given parasitic disease; the correct formulation of research problems; and the choice of appropriate experimental models. This link is also essential in order to verify to what extent findings obtained by the use of in vitro systems and animal models are applicable to man, keeping in mind the prevailing epidemiological conditions.

The collaboration of investigators in developed nations with those in endemic areas should not be shortsighted or opportunistic, a view frequently held in the past, when samples and data were collected in less-developed countries to be used for experiments to be done at home, without participation of local research teams. This attitude failed completely to benefit the local investigators for it denied them what they had expected to be a unique training opportunity; it is and was regarded as yet another form of cultural exploitation. It also failed to provide opportunities for the investigator from a nonendemic area to become familiar with the epidemiological and clinical features of the disease under study, an insight that can be obtained only by genuine collaboration and exchanges of experiences and by joint planning of the research study and interpretation of the findings. Through an awareness of the existing variables and limitations such collaboration results in a more appropriate experimental design, more reliable methods, and a more accurate evaluation of the results.

Furthermore the presence of trainees from endemic areas working in laboratories geographically removed from the health problems of their respective regions provides a positive educational experience for workers whose laboratory experience has been limited only to pathogenic organisms.

I would like to conclude by stating that the interdisciplinary approach has considerably modified the composition of laboratories in which parasitological research is conducted. It has become essential to interact on a regular basis with well-trained investigators in immunology, molecular biology, microbiology, and biochemistry. Such close collaboration of individual investigators and their respective laboratories results in a profitable exchange of ideas and expertise.

In my experience this is scientifically very productive and stimulating, in addition to providing ideal training opportunities. In other instances, however, parasitological research has moved completely into laboratories of immunology, pharmacology, or cell biology, and as a result earlier research goals have frequently been changed. This has also proved to be productive and at times has led to uncovering basic biological phenomena of a more general nature.

But, as in the case of collaboration with investigators from endemic areas, I feel it is necessary to maintain an effective dialogue between trained parasitologists and experts in other areas; otherwise we run the risk of returning to another kind of isolationism and elitism that can only hinder the progress of a science that is essential for bringing about better living conditions for a great many members of the human race.

DISCUSSION

PINO: Dr. Nussenzweig, what are the other most important divisions or departments for interdisciplinary studies in parasitology?

R. NUSSENZWEIG: There is a considerable amount of collaboration and interaction among departments of biochemistry, immunology, and, more recently, genetics, in terms of genetic cloning and population genetics. There are a number of other areas, but these are probably the most important.

TRAGER: Nussenzweig mentioned a world health aspect of our subject, which is something we have not been emphasizing sufficiently. It draws people to it because there is a certain element of altruism among scientists. In this connection, among sources of support for parasitological research the World Health Organization is actively supporting some fields, particularly the immunology of malaria, in which it is conducting a quite successful program. In the same vein the United States Agency for International Development (AID) has for many years been investigating the possibility of developing a vaccine for malaria. Indeed the AID was one of the few agencies to support this idea at a time when most others believed it to be infeasible.

PINO: There is little doubt that international agencies such as the World Bank and the AID recognize that a major impediment to the development process in Third World countries is the health problems of the people, not just the need for money with which to build dams. Dams and irrigation projects may, however, change the whole health picture. This aspect is extremely important.

Another element is the attitude of researchers and officials in the developing countries, who feel somewhat cheated by so-called collaborative efforts in which they don't participate; sometimes they don't even receive copies of the reports of research studies made in their countries. But that picture is changing.

HILL: One extremely important point Nussenzweig made is the role of a scientist working in a developing country. It is essential for parasitologists who go to these countries to try to train local investigators, so that, once the Americans leave, the local workers are able to continue to carry on the project.

PINO: The infrastructure is a major deterrent to that kind of situation, however, because often the indigenous workers do not have the materials or the money to continue the research.

WARREN: When we do send people overseas—and I have had a great deal of experience with this in the medical area—we should not do it prematurely. Some young people would like us to act as their travel agents and give them an opportunity to see the developing world. In order to help them start to embark on a career, however, it is very important to make sure that researchers are seriously interested in their field.

HILL: The veterinary school of Colorado State University is switching to the systems approach in teaching, and several other schools are doing the same thing. In terms of teaching parasitology with this approach, the parasitologists are getting squeezed out of more and more of their teaching time, as Burridge pointed out. At Colorado State that time is being taken over by clinicians, and the time allotted for the training of parasitologists is being reduced. I wonder if there are any comments on what can be done to change that. Most deans feel strongly about maintaining the systems approach. If you don't like it, that is unfortunate for you.

CHERNIN: I can't give general answers, but I can tell you what happened at Harvard. When they tried to get us to teach malaria as part of the brain or heart sequences, or something equally ridiculous, I said no. Then I had to explain why, because they didn't know what I was talking about.

Eventually we insisted—successfully—on teaching parasitology as a separate block; the subject simply did not fit into any of the pigeonholes arranged for it. That is the way we have taught it since. You can make very cogent arguments for not teaching the way the systems people want you to.

Dr. Kreier, you mentioned Ph.D.'s and Ph.D. candidates. I didn't know veterinary schools granted Ph.D.'s. I thought they were commonly the property, as it were, of the faculty of arts and sciences.

KREIER: Not necessarily. I earned a Ph.D. at a veterinary school. It is a common procedure. In a broad sense all Ph.D.'s and masters at Ohio State are awarded by the graduate school, but the departmental faculty have broad authority to set the criteria for graduate degrees, even in the medical school.

Historically, the veterinary schools have limited their graduate programs to graduate veterinarians. In more recent years there has been a loosening of that rule. They now have people with basic arts and sciences degrees enrolled in their graduate programs.

DELAPPE: At Harvard the Ph.D. in medical sciences is supervised directly by the Faculty of Arts and Sciences in Cambridge, not by the Medical School in Boston, which supervises only the M.D. degree.

SCHWABE: There are many other patterns. The University of California, for example, follows a graduate group pattern in biochemistry, microbiology, comparative pathology, and other fields in which the veterinary, medical, and agriculture schools and other facilities participate. Biochemists, microbiologists, and pathologists may belong to those respective groups, which are responsible for the graduate programs in those disciplines. The degree is ultimately granted by the graduate division.

KLEBANOFF: I would like to speak from the point of view of a university faculty member in a clinical department whose activities include teaching, research, patient care, and administration to a variable degree. An additional activity, which in some ways is turning into the tail that wags the dog, is the writing of grant proposals. It is becoming a major time commitment. I think the granting agencies should be more considerate of the time involved in these tasks and realize that it is taken away from the research efforts of the people involved.

To take a specific example, the Tropical Disease Research Units, Delappe indicated that of ten proposals received in 1979, two were disapproved, seven were approved, but not funded, and one was funded. When only one of eight approved applications is funded, that represents a great deal of time spent in vain by a number of individuals, time that could have been spent in the laboratory. That is not cost-effective in terms of research productivity.

SHERMAN: One of the failings of the university scene is that although good institutions support reseach in theory, they fail to translate this into providing the wherewithal to do that research.

At the University of California, for example, a third or more of one's time should be devoted to research, but the university puts no money into the department for that purpose; it is earmarked for teaching, and the research is bootlegged.

It is an unfortunate situation because, given the availability of federal funds for research, the universities are increasingly withdrawing their support. Today some teaching and research positions are funded entirely from federal sources, so the university sees no reason to commit its resources. How one reverses this trend I do not know. It is rather sad, because when we talk about university resources we mean state funds.

PINO: As Kreier pointed out, the decision on that is partly political, based on the level of research support.

KREIER: When my dean said that anyone who wanted to do research exclusively ought to be in a research institute, he didn't mean to imply he was not strongly in favor of research; he is a very active supporter of research. What he was trying to get across was his concept that if you are in an educational institution you should integrate your research activities with the teaching function of the university, regardless of the level of teaching. Anyone who wants to conduct a research program with no training components does not belong in an educational institution.

MAY: You cannot escape the fact that in most institutions explicit allocations are made for research from general funds. In the departmental allocations at Princeton, the formula is a mixture of teaching and research functions. It is quite explicit.

DELAPPE: Certain people at the NIH, myself included, attempt to defend the awarding of indirect costs to institutions because of the rising costs of energy, maintenance of floor space, and that kind of thing. Some NIH officers believe there would be more money for grants if indirect costs were abolished. But they are necessary, even though some abuses occur. I know of one institution, for example, whose indirect costs amount to 125 percent.

MAY: There are indirect costs attributable in a private institute that are not attributable in a university; no university has so high a ratio.

DELAPPE: A constant battle has been going on for twenty years, and every year it raises its ugly head once more. You can chop it off, but it is like a Hydra.

CHERNIN: The other head is that when indirect costs are awarded, as they commonly are at universities, they go to the dean's office; the investigator never sees them.

TRAINING AND CAREER OPPORTUNITIES IN PARASITOLOGY: ENTERING THE FIELD FROM ANOTHER DISCIPLINE

Alan Sher

BACKGROUND

It has been traditional in the United States to train for a career in parasitology by obtaining a graduate degree in either public health or parasitology itself. In recent years, however, it has become increasingly common to enter careers in parasitology and tropical medicine after training in other more basic sciences. My own career may be cited as an example of this pattern—although I should emphasize that it is by no means unique; many of my colleagues have entered the field by a similar route.

I received my graduate education in immunology and molecular biology at the University of California, San Diego. My Ph.D. thesis project, carried out at the Salk Institute in La Jolla, was concerned with a problem in basic immunology—the inheritance of immunoglobulin structural specificities known as idiotypes. The study of idiotypes has since grown to be a major emphasis in contemporary immunology, and it is conceivable that if I had continued my research in that area I could have built a successful career as a basic immunologist.

My decision to change direction and enter the field of parasitology was precipitated in large part by the sociopolitical fervor of that particular period—the late 1960s. The key word among those of us in our mid-twenties at that time was "relevance," and in common with many of my ideological contemporaries I was having great difficulty relating my own essential academic interest in basic research to the great social issues of that time.

Seeking "relevance," I joined a group of similarly motivated molecular biologists and immunologists at the Salk Institute who met informally to consider new career directions. A number of different fields were discussed as possible alternatives. Many of them, such as reproductive biology, agricultural science, and cancer research, were considered and rejected as being either too distant from our own area of expertise or already inundated with skilled workers.

One day, however, one of the group's members, Donato Cioli, an immunochemist, came across an article in the *Rockefeller Foundation Quarterly* by John M. Weir entitled "The Unconquered Plague."[1] The article discussed schistosomiasis as a global health problem, and this stimulated Cioli to research the literature in the field and present us with a talk on the immunology and pharmacology of that disease. Cioli's exposition of the problem was very convincing. Here was a major disease afflicting over 200 million people, and yet we could find evidence of only some half-dozen good research laboratories working on immunological or biochemical solutions to this "great plague." It was clear from the literature, however, that an exciting beginning had been made in analyzing the immunopathology of egg granuloma formation, immune effector mechanisms responsible for acquired resistance to infection, as well as mechanisms used by the parasite in evading host immunity. Schistosomiasis appeared to present both a highly relevant and fascinating problem. Flattering ourselves, we thought that we, as basic scientists, had much to contribute to research in a field that had been traditionally dominated by parasitologists and experts in tropical medicine.

Fortunately we had the freedom to act on our convictions, and four of our group, Cioli, Paul Knopf, a molecular biologist and immunologist, Italo Caesari, a biochemist, and myself, actually embarked on research careers in schistosomiasis. Cioli returned to Rome where he established a schistosome research group at the Laboratory of Cell Biology. Similarly, Caesari returned to Caracas where he set up a schistosome biochemistry laboratory at the Venezuelan Institute of Scientific Investigation (IVIC). Supported by the Edna McConnell Clark and Rockefeller foundations, Knopf established a group working on the immunology of schistosomiasis at Brown University, where he is now professor of biomedical sciences.

As the youngest member of the group I was afforded the luxury of undergoing a training period in schistosome research. I applied for, and was awarded, a National Institutes of Health (NIH) postdoctoral fellowship to work in the laboratory of S. R. Smithers at the National Institute for Medical Research in Mill Hill, London. In Smithers and his colleagues John Clegg and Roland Terry I found three highly skilled, knowledgeable, and exuberant teachers. Not only did they provide me with a first-rate fundamental background in schistosome parasitology, but they were patient enough to listen to my own ideas about how basic immunology might relate to the infection, and they accepted my conviction that the mouse was the ideal host for studying immunity to this helminth. The Mill Hill environment was also excellent because its superb Immunology Division enabled me to remain closely informed about current developments in my original field. Since then I have found the proximity of good basic immunology laboratories to be essential for maintaining a creative atmosphere for work on immunoparasitology.

After a summer working at the Wellcome Trust Research Laboratory in Nairobi, Kenya, I returned to the United States to the Department of Pathology at Harvard Medical School, where I was to work first as a research associate and later as an assistant professor. My first years in Boston were spent continuing my apprenticeship in schistosomiasis under the guidance of Franz von Lichtenberg, an excellent teacher of the pathology of schistosomiasis, who helped focus my attention on the crucial role of the inflammatory process in determining host resistance to helminth infection.

From von Lichtenberg's group I moved into my own laboratory in the newly formed Parasite Immunology Division headed by John R. David. Again, as in London, the proximity of superb immunology laboratories and the ability to interact with these groups was an invaluable aspect of my training in Boston. This became easier as immunoparasitology emerged as a prominent area of immunological research and attracted the interest of a number of basic immunologists in the Boston community. My continued training was heavily dependent on financial support in the form of grants from the Clark Foundation and the NIH and a Career Development Award in Geographic Medicine from the Rockefeller Foundation.

OPPORTUNITIES FOR TRAINING

It is apparent, I hope, from the above "case history" that my entry into and training in parasitology occurred simultaneously with a general explosion of interest in the field within the scientific community, and among immunologists and molecular biologists in particular. This period of unlimited growth appears to have reached a plateau, and career opportunities for both trained parasitologists and individuals entering the field from other disciplines are becoming extremely competitive, as is evident from the papers by Drs. Cook and Delappe in this volume.

Investigators who cannot obtain support by working under the grant of an established scientist will have to apply for the few postdoctoral fellowships in parasitology funded by the NIH, the World Health Organization (available only to individuals from developing countries), or the Rockefeller Foundation. Given the current general shortage of research funds this financial barrier to obtaining salary support to train in parasitology may be a just and natural mechanism for limiting the number of jobless researchers in coming years. Should continued expansion of the field be thought desirable, however, and be backed by promises of continued funding, the creation of new fellowship opportunities should be given high priority.

If the limitation of salary support can be overcome, the prospective trainee should have no difficulty finding a suitable laboratory environment in which to learn parasitology. Again, it is my contention that the best situation consists of a laboratory highly skilled in the parasitology of a given organism or group of organisms, while at the same time closely affiliated with other laboratories involved in the same basic science in which the individual was originally trained. In addition to the Mill Hill and Harvard laboratories already mentioned, many other environments exist that meet the foregoing criteria.

A good example, to which I am obviously partial, is the laboratory in which I now work, the Laboratory of Parasitic Diseases (LPD) of the National Institute of Allergy and Infectious Diseases (NIAID). This group consists of approximately a dozen senior investigators studying a wide range of disease problems (Table 1) from the points of view of parasite biology, biochemistry, immunology, and epidemeology. The laboratory operates excellent field programs on malaria, filariasis, schistosomiasis and Chagas' disease in Latin America and the Far East, and has helped train numerous investigators from developing countries. Members of the LPD are encouraged to interact with other NIAID laboratories, including the fine immunology groups headed by William Paul and Richard Asofsky and the many molecular biology groups located in different branches of the institute. This is therefore a very good setting in which to train parasitologists who come to the field with basic immunology or biochemistry backgrounds. Finally, postdoctoral fellows entering the laboratory are eligible for support, not only from the extramural sources I have listed, but from intramural sources such as staff fellowships, the Fogarty International Center, and the University Exchange Program funded by the Intergovernmental Personnel Act.

CONCLUSIONS

In summary, established and convenient routes now exist by which to enter careers in parasitology after first training in

TABLE 1. PARASITIC DISEASES STUDIED AT THE LABORATORY OF PARASITIC DISEASES OF THE NATIONAL INSTITUTE OF ALLERGY AND INFECTIOUS DISEASES

Amoebiasis	Leishmaniasis
Chagas' disease	Malaria
Filariasis	Schistosomiasis
Giardiasis	Strongyloidiasis
Trichomoniasis	

other more basic sciences. The now frequent passage of researchers through these routes has helped modernize and cross-fertilize parasitology and is largely responsible for the emergence of the hybrid disciplines of biochemical-parasitology and immunoparasitology. Although this recent trend is an encouraging and positive development, there is a danger that the training of "hybrid scientists" will occur at the expense of the careers of individuals more traditionally oriented toward studying the basic biology of parasites. In designing training programs and career pathways care should therefore be taken to preserve some opportunities for the study of fundamental parasitology so that the essential concepts and skills of the science can be maintained.

NOTE

1. J. M. Weir, "The Unconquered Plague," *Rockefeller Foundation Quarterly* 2 (1969): 4–23.

TRAINING AND CAREER OPPORTUNITIES IN PARASITOLOGY: AN "IDEAL" TRAINING PROGRAM

*A. Dean Befus**

As a parasitologist I am excited by the high level of interest in tropical diseases being shown internationally by such agencies as the World Health Organization (WHO), the Rockefeller Foundation, and the International Development Research Center—excited because the interest is justified and will ultimately lead to significant advances in human and animal health and to productivity in the future. I am optimistic that this current trend will enhance training and career opportunities for promising young parasitologists and other scientists. Despite my optimism, however, many of these young people will not contribute to progress in the field, in large part due to their incomplete, and thus inappropriate, training and supervision.

I intend to analyze some of these issues because I believe immediate concerted efforts by the appropriate individuals and agencies will increase the numbers, diversity, and quality of the scientists needed to attack global parasitological problems.

I shall begin to explore these complex issues by creating a hypothetical model of the "ideal" training program for a research career in the basic science aspects of parasitology, and by considering, with the obvious bias of my personal experiences

* I am grateful to Ms. N. Lyons for her excellent assistance in preparing this manuscript.

and observations, how closely current realities of training approach this "ideal." I will present an optimistic view of career opportunities and future prospects in parasitology, which may seem extreme, and thus attempt to identify some current problems and pose possible solutions that in my opinion will enhance and improve the outlook for bright, enthusiastic students in parasitology.

A model of an ideal training program for a research career in the discipline has many subjective components, and I would like to discuss a few that I feel are relevant (Table 1). The model assumes that the starting point is a talented and enthusiastic individual. Without compromise then, the first requirement is training of high quality in critical analysis and the scientific approach. These essential characteristics of a scientist are undoubtedly developed even prior to a formal education and constitute an ongoing process, which from my experience depends most heavily on keen and rigorous supervision at the undergraduate, graduate, and postdoctoral levels. This type of supervision should be encouraged and students should be directed to seek it out.

A second fundamental component of the model requires a knowledge of and expertise in a basic science specialty such as immunology, cell physiology, or pharmacology. This proficiency can then be applied, extended, and/or modified for investigations of parasitic organisms. This requirement stems from the conclusion that work in high-quality parasitology is performed at the frontiers of basic science, and that the greatest diversity of opportunities for parasitologists does not necessarily lie in the endogenous worth of parasite models, but in their

TABLE 1. REQUIREMENTS FOR AN "IDEAL" TRAINING PROGRAM IN PARASITOLOGY

High-quality training in critical analysis and the scientific approach
Knowledge of and expertise in a basic science specialty
An understanding of parasite biology
High-quality skills in communication
First-hand knowledge of the work of other investigators
Work in a large group (team) environment
Appreciation of the relevance of parasites to the health and productivity of animals and
    humans
Flexible approach to technological advances

contributions to science and medicine. A major factor that contributed to my current faculty appointment, for example, was not my parasitology per se, but the fact that parasites provide models for the study of basic mechanisms of host resistance and fundamental aspects of the biology of mast cells. While some may say this is compromising oneself, such an opinion is pathologically restrictive; I am confident that many opportunities for parasitologists lie in the basic sciences, and that a greater basic familiarity with them will contribute to advances in parasitology.

An understanding of parasite biology must be a component of the model; the extent and nature of this knowledge is, however, debatable. Contributions to parasitology may be made with a limited understanding of the organisms involved, but to achieve at levels near one's potential the greater the knowledge of the organisms the better. Whether this familiarity evolves before, during, or after mastery of a basic science specialty may not be a crucial factor; ideally the concurrent development of parasitology and, for example, membrane physiology might be desirable. Unfortunately there have been few training environments in which parasitology and the scientific specialties have coexisted, and students have been forced to acquire this training independently rather than concurrently. Perhaps recent international commitments to the study of tropical diseases will improve this situation. Personally, my curiosity about the life history and taxonomy of parasites transformed into immunology, with a subsequent merging of both disciplines.

The acquisition of high-quality skills in communication, particularly in writing, is another stringent component of the model. Essential to any peer review system is the ability to express ideas clearly and to develop hypotheses logically.[1] Unfortunately these skills are rarely incorporated formally in a training program. Apart from limited efforts in the realm of formal courses in scientific writing, trainees must depend heavily on their innate abilities and the luck of the draw with regard to the supervision and direction they receive. I am pleased to acknowledge the never-ending efforts of my various supervisors and coworkers in this area, and I would urge all supervisors to attempt to develop these skills in their students.

In the "ideal" training program it would be beneficial to experience first-hand the relevance of parasites to the health and productivity of animals and humans. At present this is difficult for trainees to achieve, but as we are all in a continuing training program such experience can be gained at various phases of our careers. Perhaps more success is attained in efforts to keep trainees and young scientists informed about the work of other investigators; first-hand knowledge of the interests of others is essential in one's own area of specialization.

Another component of the model is the need for students to belong to a team of scientists: technologists and students united by a common theme. The productivity and cross-fertilization of ideas within such a group can be rewarding, and the team approach will become increasingly essential as technology and the interfaces between specialties evolve. A degree of financial stability is also essential for the continuation of such groups, of course, and the recent trend by some agencies toward long-term funding is constructive in this regard.

The final, yet perhaps most important, component of the model is that an individual or a group must be capable of evolving as answers are arrived at, new questions are presented, and technology advances. A scientist cannot be a slave to technology by letting the hypotheses to be tested be dictated by the techniques currently being used. Training must instill in the student a degree of confidence that provides for the mobility of ideas and approaches.

Are there mechanisms in our training programs that develop these skills? One may simply be the experience that comes with exposure to a variety of techniques, problems, and environments. Trainees should be encouraged to gain experience in a variety of problems in more than one environment, and supervisors should be discouraged from holding on to good students.

Opportunities to approach the realization of this ideal training in parasitology are available, and some individuals are getting the most out of the existing system; a disproportionately large number of trainees in parasitology or other sciences do not, however, utilize its potentials. I do not place the blame for this on the students but on their supervisors and other faculty with whom they interact. Individuals in supervisory roles can have major impacts on the career decision and training program chosen by a student, and it is their responsibility to be in continual contact with new advances in parasitology so they may direct students toward optimal educational opportunities. Moreover, as I have already indicated, education does not stop at graduation; it is a continuing process, and in the future it will be increasingly important to keep abreast of the latest developments in relevant areas.

To overstate an example, I suggest there is an ever-widening gap between scientists—and thus their students—in classical parasitology and those in biochemistry, pharmacology, immunology, and physiology. The last four disciplines are the areas of parasitology deemed to require significant scientific, and therefore financial, input in the near future. I do not wish this discourse to degenerate into a consideration of what is "relevant" science; I merely wish to state that the decisions made about what research to fund should be considered, because the careers of young investigators depend on their being trained in areas that will remain relevant, at least in the immediate future. In short I feel it is fair to say that a large number of parasitologists are pursuing research interests for which it is currently difficult to find funds, and that training students in these fields is of questionable value.

I do not wish to be critical only of classical parasitology and parasitologists, because they provide the trainee with many of the essential components of the model I have been defining. Fine training can be acquired in critical analysis and the scientific approach, for example, as can essential, high-quality skills in communication and a solid background in parasite biology. Yet students must be motivated to move from a parasitology environment to another area of basic science that can subsequently be developed and extended in parasite models. At present trainees are not adequately encouraged to acquire an extensive knowledge of basic science outside the realm of parasitology. Perhaps a constructive but revolutionary suggestion would be to persuade scientists and trainees not to present their findings at parasitology meetings, but at meetings of basic science specialties. This might weaken the bonds that link parasitology societies, but it would permit the quality of the parasitological work to be evaluated by individuals well-versed in the concepts and technologies of the specialty field.

A theme central to my criticism of the present system is its poor use of existing resources. For example, despite extensive expertise in parasitology there appears to be limited interaction between classical parasitologists and skilled individuals whose work is deemed relevant by many funding agencies. Perhaps some increase in efficient resource utilization could come from the more formal publication of the research aims and interests of parasitologists in centers in different geographic regions. Colleagues studying parasitic diseases and infection might then find others with similar concerns and useful expertise more easily than they can at present.

Another serious problem lies in the area of the use of existing knowledge for the control and treatment of parasitic diseases. For example, E. J. Eastman, A. P. Douglas, and A. J. Watson demonstrated that the time it takes to diagnose giardiasis in both adults and children is often excessive, and as a result there are many extensive, unnecessary investigations, at a considerable cost to the patient.[2] Such difficulties may become profound as: 1) tropical diseases become more common in temperate areas, largely due to people traveling more frequently; and 2) increasing demands for time in the curriculum further restrict the teaching of parasitology and tropical medicine in medical school.

A similar example can be found in veterinary medicine, where there are major gaps between knowledge of the role of range management in the control of parasitic infection, anthelmintic resistance, and appropriate utilization, and how this knowledge is applied in the field. To narrow these gaps it is essential that communication channels be improved among the research scientist, the practicing veterinarian, and the farmer.[3] Other lines of communication in the specialty of veterinary medicine also need to be opened up.

Another obstacle to be overcome is the restriction imposed by government drug agencies on pharmaceutical firms. When funds are forthcoming from these companies to support studies of anthelmintic efficacy in academic institutions, scientists in academia run the risk of being accused of being lackeys of the industry.

Issues such as these raise many complex considerations, and to resolve them in an equitable fashion will require an input of funds and the generation of more career and training opportunities for young scientists.

To conclude, despite the foregoing obstacles one can be optimistic about parasitology and parasitologists. Clearly international recognition is at a high point; levels of funding are improving, thanks to the support of such agencies as the WHO, the Rockefeller Foundation, and the United States Department of Agriculture; and scientists and trainees of high quality are therefore being attracted to the field.

As the quality of parasitologists improves, so does the quality of the science, and it may be predicted that positive feedback will further enhance the system. It is our responsibility to maintain the drive to upgrade the quality of the science, and, with that as an objective, increased opportunities will exist for trainees and professionals engaged in parasitological research and training.

NOTES

1. D. Apirion, "Research Funding and the Peer Review System," *Federation Proceedings* 38 (1979): 2649–50.

2. E. J. Eastman, A. P. Douglas, and A. J. Watson, "Diagnosis of *Giardia lamblia* Infection as a Cause of Diarrhoea," *Lancet* II (1976): 950–51.

3. R. Herd, "Animal Health and Public Health Aspects of Bovine Parasitism," *Journal of the American Veterinary Medicine Association* 176 (1980): 737–43.

DISCUSSION

WARREN: I would like to comment on an issue related to those Befus and Sher brought up with respect to how an older discipline interfaces with what are considered new disciplines; trying to get them together is a slow and difficult process.

When Charles C. J. Carpenter became chairman of the Department of Medicine at Western Reserve, the Rockefeller Foundation gave us a grant to set up a division dealing with tropical medicine. We discussed a name for the unit and I suggested Division of Tropical Medicine, but Chuck said "tropical medicine is a perjorative term." What he meant was that those at the frontiers of clinical investigation in this country, in fields such as cardiology and endocrinology, looked down on tropical medicine because of the quality of research in the field. So we had to decide on another name, and we came up with "geographic medicine."

When we set up this division we participated in the meetings of the American Society for Clinical Investigation and the American Association of Immunologists, as well as those of the American Society for Tropical Medicine and Hygiene. That may be a reasonable way to deal with the problem of parasitology. When people with parasitology backgrounds move into the field of immunology or molecular biology, or vice versa, it is their obligation to compete with the best in their new fields, but to also continue the ties with the original field.

HILL: The interdisciplinary program at our university works well; parasitologists are working with biochemists and molecular and cell biologists. As a result, students have many opportunities to appreciate what is occurring in several different fields. Increased visibility is very important.

KLEBANOFF: I want to take issue with what I perceive as a feeling that people outside the field have no respect for classical parasitology; it was one of my most exciting courses in medical school. The unraveling of life cycles was tremendously interesting.

What may turn students off is the feeling that parasitology is not relevant because these diseases are not seen in medical school. The course becomes an exercise in parasitology, but not in medicine. The issue is therefore a sociological one of persuading students that these diseases are relevant, that we are all one world, and that to become knowledgeable about diseases one may not see locally is in fact an appropriate way to spend one's time. I don't think students have ever looked down on parasitology; scientists may have before the application of modern advances in the basic sciences to parasitology.

BEFUS: When I began to design the model I discussed I was concerned that I might reach the conclusion that classical parasitologists are not contributing as much as one would hope to the ideal training of parasitologists. In fact I finally concluded that a very large number of those components one acquires as a classical parasitologist include an appreciation of the scientific approach; the ability to be stimulated; an opportunity to improve one's writing skills; and flexibility. All these can come from classical parasitology, but you also have to stimulate people to get into the interfaces.

CHERNIN: To follow up on what Klebanoff and others have said, I think parasitologists make the mistake of believing what people in other fields say about them. A certain amount of entirely unwarranted contempt is expressed by individuals in other fields for those in classical parasitology, and many of us carry this around as a kind of burden, accepting the theory that we are in some respects subnormal.

My second comment underscores what Befus said about the great importance he attaches to the quality of communication, particularly written communication. Undergraduates and graduates should be taught how to write about what they are doing and what they hope to do. This should be a required part of their training, not merely an exercise.

PINO: One problem is how to identify young talent coming up from a wide range of sources. They don't necessarily have to come through a medical, veterinary, or agriculture school. They could presumably be biology or zoology majors and reach the same end point.

What attracts any university student to a major field? Suppose tomorrow we found the solution to one of the major diseases, would that elevate the profile of parasitology? What are the reasons we have not been successful in recruiting more young people? Lack of funding has been put pretty much at the top of the list, yet I'm not so sure that is the real problem.

COOK: I will talk about the problems of funding later, but that does not seem to be the whole answer. Having the best people in key posts in universities is essential if we are to attract graduate students to the field. The important outcome of the MBL course, which has been referred to several times here, will probably be in the new scientists it will bring into the field. In five to ten years graduates of this course will be in tenured university positions where they will be able to recruit graduate students. If there are scientists doing good research on malaria, good gratuates will want to study that disease.

TRAGER: My experience has been at the graduate level, but I wonder whether we shouldn't pay more attention to teaching courses on parasitism, and the symbiotic relationship in general, at the undergraduate level, at least in basic biology. This kind of discipline cuts across taxonomic fields. We teach courses such as bacteriology, and, more recently, microbiology at the undergraduate level.

A well-taught course in parasitism and symbiosis offered with a modern approach, so it would include both the basic, classical material as well as a great deal of the recent work dealing with the relationship between parasite and host and the dynamic interaction, would attract students just before they are ready to go on to graduate work.

WARREN: Just for clarification, would you include bacteria, viruses, and funguses, or would you limit the course to protozoology.

TRAGER: That might be difficult; you might have to fight the microbiologists.

SHER: I don't see a problem in recruiting good people to immunoparasitology. Many individuals now working in the field represent the finest of our immunology graduate students; immunoparasitology is no longer the backwater it was years ago. The big problem is what we do with the students in terms of creating career opportunities for them.

SCHWABE: Any field of science is exciting if you have a dynamic, enthusiastic exponent of it in the classroom—such a teacher can make that enthusiasm contagious. I see no problem in visualizing how to create dynamic parasitology courses. There are, however, some problems that fields such as bacteriology and virology don't have. For example, it is not hard to maintain forty, fifty, or 100 species of bacteria and any number of viruses in the laboratory and provide a variety of learning experiences. But to maintain a number of different animal parasites is another matter. The sheer work and the cost of maintaining living protozoa and helminths in order to provide a breadth of experience in immunology, pathophysiology, and other exciting and important disciplinary areas are problems of a different order of magnitude.

SIMPSON: I teach a graduate course on "Techniques in Nucleic Acid Chemistry" for a small number of students. Each set of two students is given a new species of hemoflagellate never studied before, and the object is to isolate and characterize the kinetoplast DNA from the organisms. We have gotten new information from this course and the students are turned on. I find it exciting to do and it is something I can approach in a comparative manner and get new and interesting information.

R. NUSSENZWEIG: I see a great number of applications from highly qualified individuals who want to enter this field and I don't think there is any problem, even by the most rigorous criteria, in selecting good students. The greater problem comes with further career development for the young investigator between the postdoctoral period and the award of the first research grant. Funding is essential because it is difficult for a laboratory to operate on the basis of individual research grants.

WARREN: The purpose of the Rockefeller Foundation's Career Development Awards was to do that. Unfortunately there are too few of them.

CERAMI: The thing I am pessimistic about in terms of the future has to do with the figures Delappe showed us with regard to how much money the NIH is going to spend in the entire field of parasitology. It is very limited in terms of the funds available for research in other diseases. Cancer alone, as an example, receives something like $1 billion a year; diabetes, $150 million; and sickle cell anemia, $27 million. Parasitology receives some $16 million.

So we are dealing with very small amounts. We are going to have fiercer and fiercer competition to get that funding, and unless people trained in biochemistry and immunology, for example, write applications that come under those respective study sections and get reviewed by them, we are in for a ferocious time. I hate to think about it, because it means that instead of people working together, there is going to be a fierce tug-of-war for a limited amount of money.

The whole question of how to increase the magnitude of the support available for this area is something we have to come to grips with. It is important that we consider this when we talk to young people about their careers. Although I am very excited about parasitology and all the things that can be done, in talking with students I point out the problems they will face. I think they are aware of them.

SOURCES OF FUNDING FOR TRAINING AND RESEARCH IN PARASITOLOGY

Joseph A. Cook

This paper describes various sources of funding for training and research in parasitology. I could find only two previous papers on this subject. The first is Paul Weinstein's presidential address to the American Society of Parasitology in November 1972, in which he reviewed carefully the history of support for work in parasitology by the United States Public Health Service (USPHS).[1] I will not try to summarize Weinstein's report here, but suggest you consult it for a complete analysis of the development of support for parasitology in this country.

In January 1979 the National Academy of Sciences held a meeting on Pharmaceuticals for the Developing Countries, at which W. S. Jordan, Jr., director of the Microbiology and Infectious Diseases Program of the National Institute of Allergy and Infectious Diseases (NIAID), reported on current programs in American academic laboratories to develop agents against selected infectious diseases.[2] That paper necessarily concentrated on the important human parasitic diseases, but it also mentioned other infectious diseases of primary importance to the developing countries. My report will cover some of the same areas dealt with by Jordan; in some cases there will be less detail concerning actual research, however, as this is covered in other papers in this volume.

Recent information was collected through responses to a series of letters to government agencies, foundations, and private industries. I sought to concentrate on fiscal year 1979, which I hoped would provide some data on trends; I also asked for information that would allow a measure of the relative importance of parasitology research compared to other areas of science. Unfortunately the responses were uneven and I was therefore not able to provide any overall assessment of trends in support or percent of effort compared to research on all infections. When this information was provided by an individual organization, however, it is included.

UNITED STATES GOVERNMENT ACTIVITIES

A summary of the estimated amount spent by government agencies in 1979 is shown in Table 1. It is not surprising that the greatest support came from the USPHS. While the total of $48.8 million may seem larger than expected, it includes funding for work on animal parasites; compared with the amounts spent on cancer and heart disease, it is small indeed.

Table 2 indicates the estimated support provided by all units of the National Institutes of Health (NIH). These data were obtained through the Computer Retrieval of Information on Scientific Projects (CRISP) of the NIH Division of Research Grants. A computer search was conducted for projects indexed as being related to parasitic diseases, and the table shows the total amounts and numbers of projects for each institute. Some

TABLE 1. ESTIMATED SUPPORT OF PARASITOLOGICAL RESEARCH
BY UNITED STATES GOVERNMENT AGENCIES
FISCAL YEAR 1979
(IN THOUSANDS OF DOLLARS)

Agency or Department	Amount
Public Health Service	$16,365
Army	11,584
Agency for International Development	10,200
Agriculture	7,028
Navy	1,995
Centers for Disease Control	1,400
Gorgas Memorial Institute	227
National Science Foundation	33
Total	$48,832

TABLE 2. COMPUTER RETRIEVAL OF INFORMATION ON SCIENTIFIC PROJECTS
UNITED STATES PUBLIC HEALTH SERVICE
FISCAL YEAR 1979
(IN THOUSANDS OF DOLLARS)

Institute, Agency, or Program	Number of Projects	Amount
Allergy and Infectious Diseases	138	$10,799
Cancer	17	1,493
Research Resources Division	24	1,316
Heart, Lung, and Blood	14	738
General Medical Sciences	18	720
Eye	6	623
Arthritis, Metabolism, and Digestive Diseases	6	360
Child Health and Human Development	3	228
Cancer Treatment Division	1	80
Mental Health	1	51
Total	228	$16,408

of these allocations are for projects whose principal goals may have been outside parasitology: those listed under cancer or cancer chemotherapy, for example, may deal with parasites that infect patients in conjunction with immunosuppressive drug therapy for cancer.

In his chapter in this volume Irving P. Delappe describes the NIH programs in parasitology research. The leading institute for research in parasitology, the NIAID, has the most extensive and influential program. Its support has been consistent over many years and is especially crucial for investigator-initiated research projects.

The United States Agency for International Development (USAID) is also a major source of support for research in parasitic diseases. Because of the nature of its major concern—development—the programs of the USAID are more often devoted to control than to investigator-initiated projects. In addition, its organization by geographic region makes estimates of support in parasitology difficult. Edgar Smith, director of the USAID Malaria Research Program, estimates that from 1957 to 1972 $533 million was spent on direct bilateral assistance to control malaria; only about 0.1 percent of this amount was spent on malaria research. In 1966, as a result of the loss of confidence in insecticides, and the recognition that eradication was a remote possibility, the agency began a program of research on malaria vaccine. This program now supports work in ten laboratories and has an annual budget of approximately $2.5 million.

The USAID is also a major supporter of the United Nations Development Program (UNDP)/World Health Organization (WHO)/World Bank Special Program for Research and Training in Tropical Diseases, contributing $4 million a year to this endeavor.

In addition, the USAID supports parasitology research in connection with projects overseas. The research component of the project in West Africa to control onochoceriasis amounts to approximately $250,000 a year. Annual funding for the research efforts to control malaria and schistosomiasis in the Blue Nile Health Project of Sudan will amount to approximately $500,000. A new project beginning in Cameroun is concerned with research on and the control of the transmission of schistosomiasis in artificial lakes, rice cultivation paddies, and fish ponds. It will include a malacology laboratory and a component for training in parasitology, malacology, epidemiology, and biostatistics. The USAID contributes about $3.2 million a year to this project. Smaller projects on the control of schistosomiasis are being conducted in Liberia and Swaziland.

The United States Army program in parasitology is carried out in laboratories at the Walter Reed Institute for Medical Research and in laboratories in Malaysia, Brazil, and Kenya. Table 3 summarizes the approximate amounts spent on four

TABLE 3. ESTIMATED SUPPORT OF PARASITOLOGY RESEARCH
UNITED STATES DEPARTMENT OF THE ARMY
FISCAL YEAR 1979
(IN THOUSANDS OF DOLLARS)

Disease	Amount
Malaria	$ 8,488
Trypanosomiasis	1,624
Leishmaniasis	908
Schistosomiasis	564
Total	$11,584

major parasitic diseases in fiscal year 1979. Parasitology research constitutes over 50 percent of the total funds spent on infectious diseases by the army, indicating the importance of parasitology to its medical research effort. Malaria receives the greatest concentration of support—almost $8.5 million—which is devoted to an antimalarial drug development program; a small sum is spent on vaccine development. Of the total budget of $11,584, 46 percent is spent on extramural contract projects performed by nonmilitary scientists and laboratories and 54 percent is allocated to intramural projects; 9 percent of this total supports work in overseas laboratories where important parasitic diseases of man are endemic and a major cause of morbidity.

The United States Navy also supports work in parasitic diseases in the laboratories of the Naval Medical Research Units in Cairo and Manila; the latter maintains a detachment in Indonesia. The total amount spent in 1979 was almost $2 million, 25 percent of which funded contract projects on malaria and schistosomiasis carried out by nonmilitary scientists. The navy's primary interest has been in the immunology of schistosomiasis and malaria, with a major emphasis on vaccine development. The medical research budgets of the armed services have recently come under close scrutiny, however, and it is possible that this important source of parasitology research funds will be reduced.

The United States Department of Agriculture (USDA) has an extensive program of research in parasitology; it totaled just over $7 million in fiscal year 1979. In Table 4 this program is broken down by commodity, and the amounts include both federally supported projects and state grants for work in these areas. The total constitutes 2.6 percent of the animal research budget. Only 10 percent of the $7 million is spent on biological or vector control; 55 percent goes to vaccine development; and 35 percent to treatment.

As Weinstein has pointed out, the Office of Malaria Control in War Areas was created during World War II, and its expertise in epidemiology, field control, and diagnostic parasitology was carried over when the war ended and it was renamed the Communicable Disease Center, and later the Centers for Disease Control (CDC).[3] The CDC supports work on

TABLE 4. ESTIMATED SUPPORT OF PARASITOLOGY RESEARCH
UNITED STATES DEPARTMENT OF AGRICULTURE
FISCAL YEAR 1979
(IN THOUSANDS OF DOLLARS)

Commodity	Amount
Beef cattle	$1,700
Poultry	1,334
Swine	1,163
Dairy cattle	1,120
Sheep	800
Fish	196
Small animals	23
Other	692
Total	$7,028

chemotherapy within the Parasitic Disease Service. The centers' Bureau of Tropical Disease conducts research on malaria and other parasitic diseases at its laboratory in Chamblee, Georgia, as well as at its laboratory in San Salvador. At present the Bureau of Laboratories in Atlanta is responsible for several training courses in parasitology, particularly diagnosis, as part of its general services to public health in America.

Although it is now part of the USPHS the amount spent by the CDC in fiscal year 1979 ($1.4 million) is shown as a separate item in Table 1 because its expenditures are not reflected in a computer search of grant-making agencies of the USPHS.

The two smallest organizations engaged in parasitological research with federal government funds are the National Science Foundation (NSF) and the Gorgas Memorial Institute. The NSF supported only one project in parasitology in fiscal year 1979 at a cost of $33,000, or 0.4 percent of its budget for research in systematic biology.

The Gorgas Memorial Institute is the parent organization of the Gorgas Memorial Laboratory of Panama, a unique body with a long history of work on parasitic diseases. A joint venture of the United States and the Republic of Panama, the laboratory was established in 1928 by an act of Congress as a living memorial to Major-General William Crawford Gorgas in the form of an international center for the study of diseases of the tropics. The laboratory is important as a site for training investigators for careers in research and for giving them short exposures to tropical diseases. Supported by the United States navy, an annual course is offered in tropical medicine; more recently, a course was given in in vitro cultivation of malaria parasites. In fiscal year 1979, 58 percent of the laboratory's programs in infectious diseases was devoted to work in parasitology, including malaria, filariasis, Chagas' disease, and leishmaniasis, at a cost of $227,000 (Table 1); some of these funds came from the USAID and the United States armed forces.

It was not possible to survey all possible governmental agencies of other countries, but I did obtain information from Canada. Our neighbors to the north support parasitology research primarily through the Canadian International Development Research Center (IDRC), which spent approximately $3.3 million (Canadian) in 1979. Although some of this work takes place in Canada, most of it is carried out in developing countries where parasitic diseases are endemic—African trypanosomiasis and onchocerciasis, and the vectors of each, as well as East Coast fever and fish parasites.

The Canadian IDRC also supports the UNDP/WHO/ World Bank Special Program for Research and Training in Tropical Diseases. This is a global attempt to reverse the long-standing neglect of the five important human parasitic diseases—malaria, schistosomiasis, filariasis, trypanosomiasis, and leishmaniasis—and one bacterial disease, leprosy. In each disease category, the program takes a broad perspective, usually supporting work on immunology and the development of vaccines; basic laboratory and field studies intended to improve methods of treatment; and epidemiological research. Special working groups have been formed to study possible overlapping areas of biomedical sciences; to strengthen research programs; and to conduct social, economic, and epidemiological surveys. In addition to the almost $11 million awarded in research grants, $1.4 million was spent on training activities and $3.8 million on strengthening research capabilities in developing countries where parasitic diseases are endemic. It would be neither possible nor appropriate to attempt to describe all the activities of the program at this time.

Table 5 summarizes the research projects of the special

TABLE 5. EXPENDITURES BY THE UNITED NATIONS DEVELOPMENT PROGRAM/
WORLD HEALTH ORGANIZATION/WORLD BANK SPECIAL PROGRAM FOR
RESEARCH AND TRAINING IN TROPICAL DISEASES
FISCAL YEAR 1979
(IN THOUSANDS OF DOLLARS)

	Number of Projects	Amount
Malaria	103	$ 3,006
Schistosomiasis	65	1,766
African trypanosomiasis	36	1,766
Filariasis	52	1,562
Chagas' disease	37	613
Leishmaniasis	43	612
Biomedical sciences	16	482
Epidemiology	3	460
Biological control of vectors	27	341
Social and economic	8	163
Total	390	$10,771

program in each of the five parasitic diseases and the amounts
spent for each in fiscal year 1979, a total of $10,771,000. It is
proposed to increase the budget of the program to a level of
about $25 million a year, but this of course is contingent on
continuing donor support. Its principal sources of funds at
present, apart from the Canadian IDRC and the USAID, are
the UNDP, the World Bank, and governmental agencies of
many other countries.

PRIVATE INDUSTRY

Obtaining information from government agencies regarding their programs in parasitology is a relatively straightforward, if time-consuming, exercise. While it is clear that the
contributions of private industry to biomedical and health-
related research are great, an assessment of the size and scope
of its effort in parasitology is difficult. A complete analysis
would have entailed writing to all pharmaceutical companies—
to all members of the Pharmaceutical Manufacturers' Association, for example. Instead I focused on known parasitology
divisions of pharmaceutical industries in order to determine
the likely major companies involved and to assemble a list of
persons who might respond to queries. My assessment of the

extent of industry support of parasitology undoubtedly is incomplete—as a result of both limited time to devote to the project and limited knowledge of this field.

Nearly all of the companies contacted responded. In a few cases it was to indicate that they no longer had a research program in parasitology; more often it was an informative letter about their particular interests in parasitology and a statement that varied from one company to another, but amounted to "exact amounts spent on research in given fields are considered confidential. Release of this information might be detrimental to the company's best interests." This response is more understandable when one realizes that the time required to develop and bring a drug to market may take as long as the period of protection given by a patent. Still, several companies were not constrained by a need for secrecy.

The companies that responded are shown in Table 6, with a grid indicating their current major concerns. It indicates that a number of European countries have strong programs in parasitology. Hoechst, through its affiliate Behringwerke, does diagnostic tests for parasitic diseases. Its work on the chemotherapy of malaria is supported in part by the Ministry of Research and Technology of the Federal Republic of Germany. In addition, Hoechst cooperates with the WHO in a screening program for antifilarial compounds. Rhone-Poulenc of France has an extensive program in parasitology. Most of its emphasis at present is on the chemotherapy of schistosomiasis, but other parasites, human and animal, are included. Janssen Pharmaceutica has a major interest in parasitology; in fact over half its work in infectious diseases is devoted to parasitology. Although the Wellcome Foundation Limited of the United Kingdom provided information on funding, few details were available on its current major interests. CIBA-Geigy continues to have an extensive program in veterinary and human parasites, both in Switzerland and at its research center in Bombay; it receives some support through the UNDP/WHO/World Bank Special Program for Research and Training in Tropical Diseases. Bayer has a continuing program of unknown dimensions, but recently it has concentrated on the development of a new antischistosomal compound, praziquantel. The estimated

TABLE 6. SUPPORT OF RESEARCH IN PARASITOLOGY BY THE PHARMACEUTICAL INDUSTRY

Company	Malaria	Schisto-somiasis	Fila-riasis	African Trypano-somiasis	Chagas' Disease	Leish-maniasis	Tricho-moniasis	Ame-biasis	Cicci-diosis	Helmin-thiasis
Foreign										
Hoechst	X		X		X					
Rhone-Poulenc Santé	X	X	X				X	X	X	X
Janssen Pharmaceutica	X	X	X	X	X	X	X	X	X	X
Wellcome Foundation Limited (United Kingdom)		X		X						X
CIBA-Geigy	X	X	X			X	X	X		
Bayer		X	X			X				X
Domestic										
Schering	X							X	X	X
Eli Lilly									X	X
Burroughs-Wellcome[1] (United States)	X			X		X				
Smith Kline										X
Warner-Lambert	X		X	X		X				
Merck Sharp and Dohme									X	X

Note: 1) A subsidiary of the Wellcome Foundation Limited.

cost of developing this drug, which is not yet on the market, is $21 million.

Companies in the United States generally have smaller programs, both in scope and in funding. Schering, Eli Lilly, Smith Kline Animal Health Products, and Merck Sharp and Dohme concentrate on drugs for veterinary use. The Burroughs-Wellcome Company's Research Laboratories in North Carolina are engaged in research on the biochemical parasitology of trypanosomes, leishmania, and malaria. The program at Warner-Lambert is largely supported by contracts and grants from the United States army and the WHO for work on malaria, filariasis, trypanosomiasis, and leishmaniasis.

Table 7 lists the seven companies that provided data on funding for parasitology research; in this small sample, European firms seemed less reticent than their American counterparts. The estimated total for fiscal year 1979 was $22,327,000. This amount, presumably for targeted research, equals that spent by the USPHS and the USDA for what in most cases is free-ranging, investigator-initiated research. Since many companies feel it would be damaging to release data on funding in any particular area, I have summarized only the information from those firms that were kind enough to supply it.

FOUNDATIONS

A number of foundations support research in parasitology as a part of a larger or more general program. The Macy Foundation, a cosponsor of this conference, convened a meeting in New Orleans in 1973 on Teaching Tropical Medicine,

TABLE 7. PHARMACEUTICAL COMPANIES PROVIDING DATA ON FUNDING
FOR RESEARCH IN PARASITOLOGY
FISCAL YEAR 1979

Burroughs-Wellcome (United States)
Hoechst
Janssen Pharmaceutica
Merck Sharp and Dohme
Rhone-Poulenc Santé
Warner-Lambert
Wellcome Foundation Limited (United Kingdom)

Total estimated expenditure: $22,327,000

and in 1980–81 two investigators receiving support for a sabbatical year through the Macy Scholars Program have a primary interest in parasitic disease. The W. K. Kellogg Foundation conducts a Latin-American Fellowship Program, and parasitology or tropical medicine is one of the areas of study.

Three foundations have major programs in parasitology (Table 8). In 1979 the Rockefeller Foundation spent $2.5 million, largely in support of a network of twelve university-related laboratories devoted to "great neglected diseases." Needless to say, parasitic diseases are among these, but they are not the only neglected diseases; a slightly greater amount, $2.8 million, was spent on other infectious diseases. The Rockefeller Foundation makes small grants-in-aid of $25,000, of which sixteen were related to parasitology in 1979. In that year, also, Career Development Awards in the "neglected diseases"—an amount of $50,000 a year for five years—were made to five scientists. In 1980 the Rockefeller Foundation and the Edna McConnell Clark Foundation supported the development of a new ten-week summer course on "The Biology of Parasitism" at the Marine Biological Laboratory in Woods Hole—an important addition to training in parasitology research.

The Clark Foundation's Program in Tropical Disease Research is currently devoted to the control of schistosomiasis. In fiscal year 1979 almost $2.2 million was spent on investigator-initiated projects in the three major program areas: immunology and vaccine development; biochemistry and drug development; and epidemiology and control. The division of emphasis up to now has been roughly 45 percent, 31 percent, and 24 percent, respectively. In addition, the Clark Foundation has sponsored workshops to sharpen our focus in selected scientific

TABLE 8. PRIVATE FOUNDATION SUPPORT OF RESEARCH IN PARASITOLOGY
FISCAL YEAR 1979
(IN THOUSANDS OF DOLLARS)

	Amount
Rockefeller Foundation	$2,515
Edna McConnell Clark Foundation	2,176
Wellcome Trust	1,301
Total	$5,992

areas. The published report of these workshops, *Schistosomiasis Research—The Strategic Plan* by Donald B. Hoffman, is believed to be a useful guide for the entire field as well as for the foundation.[4]

The Wellcome Trust's long and continuing interest in tropical medicine was fostered by its founder, Sir Henry Wellcome. In fact at the opening session of its 1980–81 meeting the Royal Society of Tropical Medicine celebrated the centenary of the partnership of Henry Wellcome and Silas Burroughs. In recent years the trust's support of tropical medicine units in Nairobi, Belem, Bahia, Bangkok, and Vellore has been a very significant contribution to research and training in parasitology. In fiscal year 1979 the trust funded individual research projects in malaria, schistosomiasis, and filariasis. Support for overseas laboratories in 1981 will amount to about $1.1 million. The focus of these programs is not exclusively on parasitology, but on leishmaniasis, Chagas' disease, malaria, and schistosomiasis.

Separate support for training in parasitology is rare. The USPHS through the NIAID is the most important source of funds in the United States. Delappe describes this program in his chapter. It is impossible to make an estimate of the amount of individual grants for graduate assistants and postdoctoral associates, yet these are important sources of funds for training.

In Europe the German pharmaceutical industry supports the Robert Koch Stiftung (foundation) which conducts a training program in parasitology. In 1981 the Burroughs-Wellcome Fund in the United States is planning to initiate a program of training in the molecular and biomedical aspects of parasitic diseases.

How does support of research in parasitology compare with other areas of biomedical research? Not well, as most already realize. The budget of the National Cancer Institute is $997 million, just $3 million short of $1 billion; that of the National Heart, Lung, and Blood Institute is $527 million; in contrast, the budget of the NIAID, which must support research on infectious diseases as well as on allergy and immunology, is only $214 million. From a total budget of $164 million in 1979 the American Cancer Society, a voluntary organization, spent $49 million on cancer research—almost five times the sum the NIAID is able to spend on parasitology research (Table 2). Another public voluntary organization, the Multiple Sclerosis Society, spends $6 million a year on research.

On the international level it is estimated that the annual budget for research on population is $240 million. The budget of the Consultative Group on International Agricultural Research* is $120 million a year—compare those figures with the budget of UNDP/WHO/World Bank program in tropical diseases, which is struggling to reach $25 million.

A recent article in the *New England Journal of Medicine* by a committee of the Institute of Medicine reported on research opportunities in alcoholism—clearly an area that has been neglected and deserves more attention. The committee stated that each year $209 is spent on research for each case of cancer, $8 for each case of cardiovascular disease, and only $1 for each alcohol-related case.[5] According to data I have been able to assemble, comparable figures for parasitic diseases would be low indeed.

Tables 9 and 10 show the sources and amounts of support for research on schistosomiasis and malaria, respectively. Based on the WHO estimates of the number of persons harboring these parasites, 4.5 cents are spent annually for each case of schistosomiasis and 12 cents for each clinical case of malaria; it

TABLE 9. SOURCES OF SUPPORT FOR RESEARCH IN SCHISTOSOMIASIS
FISCAL YEAR 1979
(IN THOUSANDS OF DOLLARS, ESTIMATED)

Source	Amount
United States Public Health Service	$2,286
Edna McConnell Clark Foundation	2,176
World Health Organization	1,767
Rockefeller Foundation	950
United States Department of the Army	564
United States Department of the Navy	560
Wellcome Trust	270
Total	$8,573

---

* This group, active in thirty-three countries, is a joint effort by the United Nations Food and Agriculture Organization, the World Bank, and philanthropies such as the Rockefeller and Ford foundations and the Leverhulme Trust of the United Kingdom.

TABLE 10. SOURCES OF SUPPORT FOR RESEARCH IN MALARIA
FISCAL YEAR 1979
(IN THOUSANDS OF DOLLARS, ESTIMATED)

Source	Amount
United States Department of the Army	$ 8,488
World Health Organization	3,005
United States Agency for International Development	2,500
United States Public Health Service	1,937
United States Department of the Navy	636
Wellcome Trust	370
Rockefeller Foundation	325
Total	$17,261

might be 2 cents or less for each person infected with malaria parasites.

Clearly there should not be a reduction in research on cancer, heart disease, or alcoholism; they deserve increased support. At the same time we are at a point where advances in molecular biology and immunology can be exploited through parasitology research. There are opportunities now to make real strides in understanding these diseases and designing better means to control them.

I hope some of the information in this paper will be useful as the compelling research opportunities in parasitology are presented to various individuals and organizations responsible for making decisions on funding for biomedical research.

NOTES

1. P. Weinstein, "Parasitology and the United States Public Health Service: A Relation of a Science and Government," *Journal of Parasitology* 59 (1973): 3–14.

2. W. S. Jordan, Jr., "Current Programs in United States Academic Laboratories for Development of Agents Against Selected Infectious Diseases," in *Pharmaceuticals for the Developing Countries* (Washington: National Academy of Sciences, 1979).

3. Weinstein, "Parasitology" (see note 1).

4. Donald B. Hoffman, *Schistosomiasis Research—The Strategic Plan* (New York: Edna McConnell Clark Foundation, 1979).

5. L. Aronow, E. O. Nightingale, and B. Filner, "Special Report. Research Opportunities in Alcoholism," *New England Journal of Medicine* 303 (1980): 595–96.

DISCUSSION

BOWERS: Dr. Cook, why did the Clark Foundation decide to support research in schistosomiasis?

COOK: The decision evolved over a period of time. Initially the Board of Trustees decided there would be four programs, three domestic and one international, the latter called the "Developing World Program." The foundation's president and trustees have always felt that the programs should be carefully focused, so that the relatively small amount of money available would have a real impact.

The "Developing World Program" examined a number of opportunities such as disaster relief and economic support for handicraft industries, but found that tropical disease research was a neglected and underfunded area with great promise of accomplishment with the funds available—approximately $2 to $2.5 million a year. Donald B. Hoffman was responsible for the evolution of this program and for the construction of a strategic plan for research on schistosomiasis, which is constantly being updated with the help of scientists within and outside the tropical disease research community.

WARREN: Cook's review of the financial situation of parasitology is the best I have ever seen. It is particularly important because of the report of the Bourne Committee, which was set up to find out how much money the government was spending on international health. The report concluded that it was giving substantial support, and many people felt it was too much and that funding should be curtailed. I believe the Jordan Report given at the meeting on Pharmaceuticals for the Developing Countries also tended to inflate the resources available. Cook's figures are more realistic.

One of the problems of funding research on parasitic diseases in the United States is that decisions are based on whether the diseases represent significant problems in this country. Basically the NIH cannot fund projects of global importance if they are not of concern here. Acting on the recommendations of Senators Jacob J. Javits and Edward M. Kennedy the government has been making attempts recently to develop international health skills.

Much to my surprise, however, according to a report of the National Council on International Health, when some of the key international health experts within the government testified before Congress on these proposals they did everything they could to scuttle them. I don't understand that psychology.

PINO: It is virtually impossible to find total government expenditures for any particular subject, as there is no coding or indexing for different research areas. You find projects in agriculture or in medicine being supported by almost every branch of government, but because there is no centralized index it is difficult to find the total figure on sources of support. There is a definite need here because funding for some areas is falling into cracks.

SHER: Dr. Cook, what do you see as the future of funding for research in the next several years? We need some idea of this in order to plan responsibly the extent to which growth should be stimulated.

COOK: From my perception of the economy I can't see how funding for research will expand in the near future. It has been proposed that an Institute for Scientific and Technological Cooperation be created, modeled after the Canadian IDRC, which would be supported by funds from the AID. This agency would make grants for research on diseases that are not of significant public health consequence in the United States. Unfortunately the bill has not received congressional approval. More directly it would be helpful if Congress would simply recognize in a public law that all "tropical" diseases are of importance to the American population and that research on diseases of international public health significance should be conducted. Federal laboratories and granting agencies need that mandate to conduct research on all diseases.

CHERNIN: There is beginning to be what are known as associate programs or partnerships between universities and industry. Money may begin to flow to the universities through that mechanism.

WARREN: The Dow Chemical Company and Harvard now have such a joint program, as have Hoechst and the Massachusetts General Hospital.

V. NUSSENSWEIG The new companies involved in genetic engineering are also investing in parasitology, particularly in vaccine development.

Dr. Victor Nussenzweig, New York University, and Dr. Eli Chernin, Harvard School of Public Health.

Dr. Nadia Nogueira, Rockefeller University, Dr. Seymour J. Klebanoff, University of Washington, Dr. Paul C. Beaver, Tulane School of Public Health and Tropical Medicine, and Dr. Anthony Cerami, Rockefeller University.

Dr. Michael J. Burridge, University of Florida, and Dr. Julius P. Kreier, Ohio State University.

Dr. Paul P. Weinstein, University of Notre Dame, and Dr. Alan Sher, Harvard Medical School.

140

Dr. Ruth S. Nussenzweig, New York University, and Dr. Kenneth S. Warren, Rockefeller Foundation.

Dr. Irwin W. Sherman, University of California, Riverside, Dr. Larry Simpson, University of California, Los Angeles, and Dr. George C. Hill, Colorado State University.

THE PRESENT STATUS OF
THE PARASITOLOGY LITERATURE

Kenneth S. Warren

When the status of a scientific field is considered, emphasis is placed on teaching and research; rarely is the state of the literature in that discipline examined, even though the literature, both quantitatively and qualitatively, is the mirror of accomplishments in any given scientific field. One reason for this situation may be that few bench investigators are familiar with the field of information science.

My own involvement with information science began accidentally, but it has become one of my major interests. One of the reasons I went to Western Reserve School of Medicine in 1963 was that I planned to keep a personal bibliography on schistosomiasis, and the Department of Preventive Medicine had one of those newfangled Xerox machines. When I began to use the Xerox to an inordinate degree the professor suggested I talk to Vaun A. Newill, a colleague who was working on a computerized annual bibliography of diabetes. I immediately became involved in this new technology, and in 1967 we produced a comprehensive computerized bibliography of the literature on schistosomiasis from 1852 to 1962.[1] This was followed by a bibliography of the best papers in the field up to 1972, selected by a board of forty-seven experts.[2] Both endeavors were repeated at the request of the Edna McConnell Clark Foundation, bringing the bibliographies up through 1975.[3,4]

Thus schistosomiasis became the most extensively and intensively bibliographed of all the biomedical literature, and this in turn offered an unusual opportunity for bibliometric analysis. Again I followed in Newill's footsteps and began to work with William Goffman, a mathematician and dean of the School of Library Science. This eventually led to my appointment as a professor of library science, and, recently, to writing a book with Goffman on a selective approach to scientific information analysis.[5]

At the Rockefeller Foundation we have now developed a program called "Coping with Biomedical and Health Information" that has among other things resulted in three conferences and three sets of working papers published by the foundation: *Coping with the Information Explosion, Selective Medical Libraries for the Developing World,* and *Research in Selective Information Systems.* *

Recently I have begun using this material about information systems as an analytical tool at scientific meetings. In 1979 at a conference celebrating the centennial of Paul Ehrlich's discovery of the eosinophil I summarized the proceedings using network analysis.[6] This demonstrated how all the scientists who presented papers at the meeting had been working together; it may even have stimulated an awareness of the collaborative process, thus leading to further collaboration.

At this conference I would like to apply some tools of information science to an investigation of the parasitology literature. First let us begin with the parasitology journals of the world as recorded in the 1980 Index of the National Library of Medicine Serial Titles, which lists approximately 20,000. Forty-five of the journals are indexed under a variety of terms for parasitology and parasites (Table 1); in addition, it lists nine under helminthology and protozoology and another forty under tropical medicine. For the purposes of this paper, however, we will concentrate on the eight major English-language parasitology journals that were founded chronologically in the order shown in Table 2.

Relative to the concerns mentioned early in my discussion of Beaver's paper—What Is Parasitology?—both the concept of

---

* Copies of all three publications are available free of charge from the foundation.

TABLE 1. PARASITOLOGY JOURNALS OF THE WORLD LISTED IN THE
1980 INDEX OF THE NATIONAL LIBRARY OF MEDICINE SERIAL TITLES

A. HELMINTHOLOGY[1]
1. Helminthologia
2. Indian Journal of Helminthology
3. Journal of Helminthology
4. Proceedings of the Helminthological Society of Washington
5. Studia Helminthologica
B. PROTOZOOLOGY
1. Acta Protozoologica
2. Journal of Protozoology
3. Protozoologiia
4. Protozoology
C. PARASITOLOGY
1. Acta Medica Italica di Malattie Infettive e Parassitarie
2. Acta Parasitologica Iugoslavica
3. Acta Parasitologia Lituanica
4. Acta Parasitologica Polonica
5. Angewandte Parasitologie
6. Annales der Belgische Verenigingen voor Tropische Geneeskund, voor Parasitologie
7. Annales de Parasitologie Humaine et Comparée
8. Annales des Sociétés Belges de Médecine Tropicale, de Parasitologie, et de Mycologie
9. Annals of Tropical Medicine and Parasitology
10. Archivio Italiano di Scienze Mediche Tropicali e di Parassitologia
11. Archivos Venezolanos de Medicina Tropical y Parasitologia Medica
12. Boletin Chileno de Parasitologia
13. Ceskoslovenska Parasitologie
14. Dynamic Aspects of Host-Parasite Relationships
15. Experimental Parasitology
16. Folia Parasitologica
17. Indian Journal of Parasitology
18. International Journal of Parasitology
19. Japanese Journal of Parasitology
20. Journal of Parasitology
21. Korean Journal of Parasitology
22. Meditsinskaia Parazitologiia i Parazitarnye Bolezni
23. Parasite Immunology
24. Parasitica
25. Parasitologia Hungarica
26. Parasitologische Schriftenreihe
27. Parasitology
28. Parassitologia
29. Parazitologicheskii Sbornik
30. Parazitologiia
31. Parazity Zhivotnykh i Rastenii
32. Problemy Parazitologii
33. Revista de la Catedra de Microbiologia y Parasitologia
34. Revista Iberica de Parasitologia

35. Revista Latinoamericana de Microbiologia y Parasitologia
36. Revista de Medicina Veterinaria y Parasitologia
37. Rivista di Parassitologia
38. Symposia, British Society for Parasitology
39. Systematic Parasitology
40. Veterinary Parasitology
41. Wiadomosci Parazytologiczne
42. Zeitschrift für Infektionskrankheiten, Parasitaere Krankheiten und Hygiene der Haustiere
43. Zeitschrift für Parasitenkunde
44. Zeitschrift für Tropenmedizin und Parasitologie
45. Zentralblatt für Bakteriologie, Parasitenkunde, Infektionskrankheiten und Hygiene 1 Abt Medizinisch Hygienische Bakteriologie, Virusforschung und Parasitologie

NOTE: 1) Serial index heading.

the discipline and its subject matter have evolved as expressed in manifestos presented in the initial issues of the major journals in the field. The introductory statement in the first issue of *Parasitology* is vague: it mentions as appropriate subject matter disease-transmitting insects, malaria, trypanosomiasis, spirochaetoses, piroplasmosis, and plague, as well as parasitic worms. The *Journal of Parasitology,* which appeared somewhat later, presents a more circumscribed picture in its "announcement," as it mentions animal parasites. It goes on to state that: "Emphasis will be laid on the morphology, life history and biology of zooparasites, and the relations of animals to disease." The third issue of the first volume of *Experimental Parasitology* in 1952 states that it

> will publish papers dealing with all experimental approaches to problems in the field of parasitology. Contributions that involve experimental physiological, metabolic, biochemical, nutritional, and chemotherapeutic problems of parasites and host-parasite relationships will be emphasized.

TABLE 2. MAJOR ENGLISH-LANGUAGE PARASITOLOGY JOURNALS AND YEAR OF INCEPTION

Parasitology	1908
Journal of Parasitology	1914
Journal of Helminthology	1923
Experimental Parasitology	1951
Journal of Protozoology	1954
International Journal of Parasitology	1971
Parasite Immunology	1979
Molecular and Biochemical Parasitology	1980

The initial issue of the *International Journal of Parasitology* does not specify subject matter, but states that its aim

> is not only to serve as a vehicle for the publication of research papers but also to act as a means of communication between parasitologists at an international level.

In a recent issue of *Science*, John M. Ziman, a physicist and frequent writer on information science, points out that

> the traditional process of evolution [of scientific literature] by speciation, [that is] differentiation and subspecialization of journals . . . may offer new outlets for numerous eager authors in the newly defined field of research.[7]

The journals covering the fields of helminthology and protozoology, as they appeared, respectively, in 1923 and 1954, do not seem to have followed a clearcut temporal pattern. Two new journals may, however, fulfill Ziman's criteria: *Parasite Immunology* began publishing in 1979 because "the last decade has witnessed a great upsurge of interest in the immunology of parasitic infections." *Molecular and Biochemical Parasitology* arrived in 1980 as "a medium for the rapid publication of investigations of the molecular biology and biochemistry of parasitic protozoa and helminths."

A quantitative assessment of papers on parasitology in nineteen major English-language journals in 1979 is shown in Table 3.* The four general parasitology journals included 429 papers, the specialty journals 106, and the three major tropical medicine journals 280. It is of interest that none of these journals is a monthly; they appear at less frequent intervals. Of the two major weekly general medicine journals in the world, the *New England Journal of Medicine* published only three papers on parasitology in 1979—one original paper on toxoplasmosis and two reviews on filariasis and leishmaniasis as part of a special series; *Lancet*, the other, produced twenty-two papers, many of which were on malaria. The *Journal of Infectious Diseases* and the *Journal of Experimental Medicine* included, respectively, nine and

---

* This and much of the subsequent information in this paper was prepared with the aid of Eli Chernin, professor of parasitology, Harvard School of Public Health, and book editor, the *New England Journal of Medicine.*

TABLE 3. PAPERS ON PARASITOLOGY IN NINETEEN ENGLISH-LANGUAGE JOURNALS, 1979

GENERAL PARASITOLOGY		
Parasitology	66	
Journal of Parasitology	195	
Experimental Parasitology	94	
International Journal of Parasitology	74	429
SPECIALTY PARASITOLOGY		
Journal of Protozoology	57	
Journal of Helminthology	49	106
TROPICAL MEDICINE		
Transactions of the Royal Society of Tropical Medicine and Hygiene	114	
American Journal of Tropical Medicine and Hygiene	99	
Annals of Tropical Medicine and Parasitology	67	280
GENERAL MEDICINE		
New England Journal of Medicine	3	
Lancet	22	25
SPECIALTY MEDICINE		
Journal of Infectious Diseases	9	
Journal of Experimental Medicine	7	16
VETERINARY MEDICINE		
American Journal of Veterinary Medicine	33	
Journal of the American Veterinary Society	28	61
GENERAL SCIENCE		
Science	2	
Nature	15	17
SPECIALTY SCIENCE		
Journal of Immunology	33	
Journal of Biological Chemistry	5	38
Total		972

seven papers. Two major veterinary journals had sixty-one papers on parasitology. Of the two great weekly general science journals, *Nature* published fifteen papers on parasitology (1.1 percent of its output); and *Science* two (0.3 percent of its output); most of the papers in these journals were concerned with protozoa. *The Journal of Biological Chemistry* published five papers, and the *Journal of Immunology* thirty-three, twenty-three of them on *Schistosoma mansoni*. Thus a total of 972 papers were published in all these journals in 1979. Compare this with the 960 papers that appeared in 1979 in the *Journal of Immunology* alone. Quantitatively, therefore, the field of parasitology does not appear to be flourishing in major world journals.

The general subject matter of the papers in the four major parasitology journals, as assessed by Chernin, is shown in Table 4. They average 52 percent in the more traditional subject areas of taxonomy, morphology, ecology, epidemiology, and in vitro cultivation, and 41 percent in the newer fields of immunology, biochemistry, physiology, and molecular biology. *The Journal of Parasitology* has the lowest score of 29 percent in the latter, while *Experimental Parasitology* has 68 percent. There is little question that, in spite of complaints of commercialism, *Experimental Parasitology* has consistently provided a forum for these areas of the discipline.

Sixty percent of the articles in the *Journal of Protozoology* were in the more traditional fields and 36 percent were in

TABLE 4. DISTRIBUTION OF SUBJECT MATTER IN FOUR MAJOR PARASITOLOGY JOURNALS, 1979
(PERCENT)

Subjects	Parasitology	Journal of Parasitology	Experimental Parasitology	International Journal of Parasitology	Total
Taxonomy, morphology, ecology, epidemiology, in vitro studies	55	66	24	49	52
Clinical, pathology, diagnosis, pharmacology, treatment	8	5	7	11	7
Immunology, biochemistry, physiology, molecular biology	38	29	68	40	41

immunology, biochemistry, physiology, and molecular biology; the breakdown for the *Journal of Helminthology* was 65 percent and 16 percent, respectively.

The three journals specializing in tropical medicine averaged 37 percent for traditional parasitological fields; 35 percent for clinical areas; and 27 percent for immunology, biochemistry, physiology, and molecular biology (Table 5).

On the basis of the data on the output of these journals the question arises as to whether parasitologists are involved enough in the more modern fields of biology; further qualitative questions would be difficult to answer. For example, are the papers on immunology in the *Journal of Parasitology* comparable to those in the *Journal of Immunology?* Are the nine papers on biochemistry and physiology in *Parasitology* comparable to those in the *Journal of Biological Chemistry?* Comparability refers not just to subject matter but to experimental design and execution. More extensive studies involving citation analysis may enable us to approach some of these questions in the future.

The quality of the journals, as well as that of individual papers, may be evaluated by citation analysis; in fact because the journals cover broader areas their assessments are simpler

TABLE 5. DISTRIBUTION OF SUBJECT MATTER IN THREE MAJOR
TROPICAL MEDICINE JOURNALS, 1979
(PERCENT)

Subjects	Transactions of the Royal Society of Tropical Medicine and Hygiene	Annals of Tropical Medicine and Parasitology	American Journal of Tropical Medicine and Hygiene	Total
Taxonomy, morphology, ecology, epidemiology, in vitro studies	36	49	31	37
Clinical, pathology, diagnosis, pharmacology, treatment	36	33	36	35
Immunology, biochemistry, physiology, molecular biology	28	19	32	27

to interpret. The best estimate is provided by the impact factor based on the number of citations of articles in a particular journal, divided by the total number of articles it publishes. The 1978 issue of *Journal Citation Reports* published by the Institute for Scientific Information (ISI) in Philadelphia covered 3,231 journals from both the Science Citation Index and the Social Sciences Citation Index of the ISI for the years 1976 and 1977 and provided the ranking and impact factors for the major English-language parasitology and tropical medicine journals (Table 6).

Before closing, it might be well to turn back to where I began. This field has been blessed with a unique bibliographic compendium, the *Index-Catalogue of Medical and Veterinary Parasitology* published in Beltsville, Maryland. This system was started at the end of the nineteenth century and has been meticulously maintained ever since. It was the single most comprehensive literature source for our first schistosomiasis bibliography. In addition to card files, eighteen volumes covering the period 1932 to 1952 have been published with annual supplements up to the present time. Although a complex indexing system makes the *Index-Catalogue* difficult to use, this major work provides a comforting sense of completeness.

In conclusion, this analysis of the parasitology literature

TABLE 6. RANK BY IMPACT FACTOR[1] OF MAJOR ENGLISH-LANGUAGE
PARASITOLOGY AND TROPICAL MEDICINE JOURNALS

Title	Rank[2]	Impact Factor
Parasitology	485	1.689
Journal of Protozoology	509	1.627
Transactions of the Royal Society of Tropical Medicine and Hygiene	544	1.556
Experimental Parasitology	622	1.413
American Journal of Tropical Medicine and Hygiene	817	1.130
International Journal of Parasitology	881	1.053
Journal of Parasitology	1,017	0.909
Annals of Tropical Medicine and Parasitology	1,107	0.825
Journal of Helminthology	1,387	0.624

NOTES: 1) Impact factor = $\dfrac{\text{number of citations in 1978}}{\text{number of articles published in 1976 and 1977}}$. 2) Rank out of 3,231 journals in the science and social science citation indexes of the institute for Scientific Information.

seems to suggest some deficiencies in its quantity and in its distribution of subject matter with respect to modern biological disciplines. Moreover, the publication profile of the field does not appear to be commensurate with the immense importance of parasitology for the well-being of mankind.

NOTES

1. K. S. Warren and V. A. Newill, *Schistosomiasis: Bibliography of the World's Literature from 1852 to 1962*, 2 vols. (Cleveland: The Press of Western Reserve University, 1967).

2. K. S. Warren, *Schistosomiasis: The Evolution of a Medical Literature. Selected Abstracts and Citations, 1852–1972* (Cambridge: Massachusetts Institute of Technology Press, 1973).

3. K. S. Warren and D. B. Hoffman, *Schistosomiasis. III: Abstracts of the complete Literature, 1963–1974* (Washington: Hemisphere Publishing Corporation, 1976).

4. D. B. Hoffman and K. S. Warren, *Schistosomiasis. IV: Condensation of the Selected Literature, 1963–1975* (Washington: Hemisphere Publishing Corporation, 1978).

5. W. Goffman and K. S. Warren, *Scientific Information Systems and the Principle of Selectivity* (New York: Praeger, 1980).

6. K. S. Warren, "A Perspective Summation," in *The Eosinophil in Health and Disease*, ed. A. A. F. Mahmoud, F. Austen, and A. S. Simon (New York: Grune and Stratton, 1980): 345–54.

7. J. M. Ziman, "The Proliferation of Scientific Literature: A Natural Process," *Science* 208 (1980): 369–71.

DISCUSSION

BEAVER: Warren has done a great service in calling to our attention the publication and distribution of articles in the field.

The Journal of Parasitology and *Experimental Parasitology* are the two main American journals that serve two different groups and serve them well. There should be no quarrel about that. Some of us don't think so highly of the *Journal of Parasitology*, however, because it doesn't publish certain kinds of papers; *Experimental Parasitology*, likewise, is not particularly meritorious in this regard. It isn't a matter of what kinds of papers are sent to the various journals. In general it is appropriate that papers should be submitted to the journals in which one may expect them to appear.

The New England Journal of Medicine has published only a few papers on parasitology research, but it shouldn't have published any because the quality was very poor. One was so bad I wrote the author and asked him to explain some points in it. He sent me the material and it was really disgraceful that this journal, which has such a high reputation, would have accepted the paper.

The kind of analysis Warren has made serves to point out that, in general, specialty journals are preferable, and that specialty journals in parasitology should publish most of the papers in that field; other journals should decline to do so unless they have reliable referee support.

WARREN: Beaver is right about the quality of papers that appear in such journals as *Lancet* and the *New England Journal.* In our analysis of the 1,738 journals that published papers on schistosomiasis from 1852 to 1962, we ranked them according to the quality of their content. We gave a minus one to every article that was not cited by any of our board of forty-seven experts. If a paper was cited ten times it was given a plus ten. To get either a very high or a very low score the journal had to publish a fairly large amount of material on schistosomiasis.

It turned out that, of the 1,738 journals, *Lancet* ranked 1,734th. It had published a great many papers on schistosomiasis that were not cited by any of the experts as being of any importance—and they cited approximately one-third of the literature.

I disagree with Beaver about the *New England Journal,* however, because I believe it and *Lancet* should publish good material on health problems of global significance. Early in 1980 the *New England Journal* published a paper by J.A. Walsh and myself called "Selective Primary Health Care: An Interim Strategy for Disease Control in Developing Countries," and we were very pleased. We received more inquiries on a general level about that article than we would ever have received had it been published elsewhere.

The point is that *Lancet* and the *New England Journal* are considered to be the two best medical journals in the world. They are read globally, and their subject matter should therefore be global in content. If good papers on parasitology are submitted to them, they should publish them because that will disseminate more information about the field than papers published in any other journals.

SCHWABE: The classification of journals in terms of their coverage of modern or classical aspects of parasitology differs a great deal depending on the approach. It is rather arbitrary to state what constitutes *categories* of modern parasitology. I will quarrel with two, particularly.

One is to put in vitro studies in the area of classical parasitology because, with respect to helminths particularly, that is certainly an area that should be considered modern parasitology.

I also quarrel about calling some epidemiological papers classical—those that deal with mathematical epidemiology in contrast to those that are merely surveys. You might have ranked some of the journals quite differently if there were more than four categories.

WARREN: I don't think there would be a significant difference.

HILL: When *Molecular and Biochemical Parasitology*, which relates to what is happening in this field, was first published it planned on six issues a year. Since then there has been a tremendous increase of interest in this area, and beginning in 1981 it will appear once a month. That is a reflection of the number of papers it receives.

TRAGER: *Infection and Immunity*, which is put out by the Society of Microbiology, publishes papers in about four months. There has been some increase in the number submitted to it, and it is a journal that people at this conference might consider sending material to.

BEAVER: I have another bone to pick. A subscription to a journal that publishes 300 papers a year costs around $30, while other journals that publish the same number of papers of about half the length cost over $200 a year. Why is that?

PINO: Many of these journals are purely commercial en-

terprises. Perhaps there is a difference in the subscription rates
of journals of professional societies and those put out by publishers for profit.

BEAVER: About two years ago an analysis was made of
scientific journals published in all languages. It was pointed out
that 26 percent of all scientific data appears in those published
for profit.

Fifty years ago there were very few such journals, and none
in this country. Two in Germany and one in the Netherlands
made a profit for their investors. I expect that today 28 percent
of the journals may be published for profit. One wonders what
the situation will be ten years from now. This is a problem for
scientific societies: Will they be able to publish a journal in
1990?

WARREN: A legitimate question might be asked in that
respect. The subscription rate depends on what the market will
bear, but why aren't these articles being published in standard
journals such as *Parasitology* and the *Journal of Parasitology*? It is
probably a failure on the part of the societies and the journals
to encourage people to submit more papers of these kinds to
them.

THE FUTURE OF PARASITOLOGY

THE FUTURE OF PARASITOLOGY:
AN OVERVIEW

Joshua Lederberg

None of my published research has been on organisms customarily called parasites. Nor is my assigned subject, "The Future of Parasitology," one that is enhanced by laboratory data. I will offer some general observations and contrast them with general wisdom on how health research is conducted. The concerted scientific attack on communicable diseases is an appropriate precedent for the problems we now address in parasitology.

The discovery in the late 1800s of bacteria as agents of disease was a revolutionary advance, albeit not a scientific revolution in T. S. Kuhn's sense. It may be the best example to date of a scientific insight becoming a reductionist foundation for practical advances of the most sweeping and important kind.

The germ theory, and the basic techniques for the recognition and isolation of various species of pathogenic microorganisms in pure culture established by Louis Pasteur, Robert Koch, and Ferdinand Cohn, gave us a sweeping scientific principle looking for the appropriate questions to which it would be a solution. As far as I am aware this principle has had more important consequences for the improvement of public health than any other concept in the history of mankind. But today we are uneasy because we cannot so easily replicate that kind of comprehensive advance for our remaining health problems such as heart disease, cancer, and schizophrenia.

The miracle drugs—the antibiotics—are the most recent and potent example of the application of these elementary principles to health problems. Nothing since has so profoundly captured the public's imagination for the justification of health research over the last twenty-five to thirty years.

The result has been increasing tension between the ultimately well-founded expectations of public constituencies, who considered our advances in infectious bacterial diseases as a prototype of the fruit of investments in health research, and the realization that we do not yet have principles of comparable power and immediate relevance to many other afflictions with which we are now concerned.

Important work has been going ahead steadily, but compared to the revolutionary advances of the first part of this century it seems to be progressing in a painfully slow fashion. At the same time we have witnessed exponential increases in government investments in health research, and the disparity between anticipated scientific discoveries and the level of funding is brought forcefully to our attention in every budget cycle of the Congress.

The introduction of the germ theory of disease led to a rapid penetration of one large set of public health problems. I hold the view that it is precisely in parasitic infection that we have the nearest analogue to that kind of opportunity. My major premise is that when our understanding of the eukaryotic agents of infection can be brought to a comparable level of depth and insight—complicated by the facts that parasites are indeed eukaryons, often intracellular in habitat, and resemble the metabolism of the host more closely than do bacterial parasites—we will see advances as sudden and as spectacular as were achieved for most of the bacterial infections.

There are of course many problems. On the social and political side is the fact that the United States is no longer a colonial and imperial power of the style that had strong motivation to find cures for diseases that wreaked havoc in tropical countries. While promoting decolonization we have shamefully neglected our human responsibility to allocate resources to help solve the health problems of the developing world, for they are not at the forefront of concern in the health statistics of our country. But even that probably unduly minimizes the impact parasitic infection will continue to have, even on the health of our own citizens and livestock.

The assemblage of scientists at this conference is testimony to the capacity to mobilize a diverse set of experiences and intellectual resources. If the material resources could be made available, with the kind of nucleation represented here, very rapid progress, a veritable new wave of research, would at this point be inevitable.

But the political obstacle is serious, and I do not understand why it is so difficult to get a broader range of public support for the kinds of questions with which we are mutually concerned. Perhaps we still have not transmitted even the bare factual message that malaria remains the world's most important disease in terms of its impact on the health, welfare, and economic and social development of vast numbers of people.

We must further not overlook the fact that even while the management of bacterial infectious disease is a paragon, our self-congratulation must be tempered, for we have by no means dealt with all the health problems in this country. We will probably see even more of them in the future, with the continued and recurrent emergence of antibiotic-resistant strains of microorganisms and, quite possibly, with the emergence of new diseases and infections caused by other aspects of the continuing evolution of that huge part of the biosphere reflected by the microbial world.

In a way, I consider our present management of bacterial infections, particularly tuberculosis and leprosy, a disgraceful testimony to the inadequacies of our scientific base. In part this is a reflection of the insufficient penetration of available scientific methodology in this field. I may be wrong, but I believe it is barbarous that we continue to use an agent such as BCG vaccine for large-scale tuberculosis immunization campaigns at a time when there is much more we should know about the biology of the organisms used in BCG. We should go much more deeply into the question of how to purify the components so we may have a choice of strains, mechanisms of pathogenesis, and so on.

Fiascos such as the one in India, as revealed in a recent World Health Organization study, are inevitable if we are to go back over sixty years for material that has not been examined from any modern, sophisticated scientific standpoint.

As for leprosy, a far more effective line of investigation could be conducted with genetic analysis of the infection. We face the serious question of how to deal with an organism that is so peculiar in its range of susceptible species. I do not know of any other area of biology that links the armadillo and the human.

To deal with either leprosy or tuberculosis requires a deeper knowledge of the behavior of our cellular defense mechanisms than we have at the present time. Getting the organisms into pure culture is only the starting point of effective science. As soon as we get that deep into human physiology we are dealing with paradigms of investigation far more intricate and difficult than, for example, growing a diphtheria organism in a test tube and finding it produces an exotoxin that will kill guinea pigs in trace amounts. It is of course no accident that, from a world public health perspective, these bacterial infections are intractable residues that bear an interesting resemblance to the parasitoses.

There is likely to be needless tension between the perspective of the experimental, laboratory-oriented investigator, on the one hand, who thinks of parasitology as an exciting, challenging playground for linking biology, life cycle, microbiology, cell biology, and developmental aspects of a fascinating group of organisms—which, by the way, happen to have public health implications—and, on the other hand, that of the sanitary, public health-oriented researcher who quite correctly points out that major advances in public health throughout the world depend far more on environmental management, sanitation, and interrupting the ecological patterns of our parasitic life cycle than on specific therapeutic measures and interventions of human physiology.

Both perspectives are absolutely indispensable for further progress in the field because, even with a global public health orientation, the vastness of the stakes is matched only by the vastness of the errors that can be made when programs are mounted without an adequate base of biological insights as to the nature and behavior of the relevant organisms—whether at the ecological or molecular level.

The concept that it is possible to eradicate almost any known disease is a treacherous one, although smallpox is an example of a disease with very special peculiarities that made it most amenable to eradication. We should keep our fingers crossed, however, because we do not know enough about the genetics and the evolution of that particular virus or its affinities with other viruses of the world. Nor do we know enough about the molecular biology of the virology of smallpox, compared to other related viruses, to be certain that outbreaks will not recur. One should be wary about the prospect of the development of a global herd that twenty or thirty years from now will not have been immunized against smallpox.

The possibility of the reevolution of the smallpox virus must be matched against the de novo emergence of a wide range of other viruses, events of perhaps comparable likelihood.

It is a matter of general concern that, as we strive for a sanitary world, we are also developing a large sensitive herd! We are, then, scarcely giving enough thought to the revolutionary biological reengineering of our own species, which is implicit in effective sanitation. Never before in our history have we had such large numbers of individuals whose immunological experiences and contacts with infectious agents have been so drastically altered because of the very success of the measures we are hoping to introduce. I have no idea what the further implications of that reengineering will be, but I do not see how it can fail to penetrate other aspects of the life cycle of the human organism in ways that will not always be what we hope for.

The question of large-scale control of any species through sanitary measures—be they vectors or parasites—cannot be undertaken thoughtfully without much deeper inquiry into the biology of the target organisms, and particularly of their capacity for future evolution.

We must be concerned about other animal hosts for the parasite and about the possibility that virulent mutants or hybrids may reemerge from the gene pool of related species. For this we already have the example of pandemic influenza, which is believed to come from viruses normally resident in birds.

Certainly those measures used in environmental control should in some way bypass the logistical problems that would be involved, for example, in providing recurrent chemotherapy to large numbers of individuals, particularly in areas of the world with poor economic conditions, inadequate transportation, and limited medical services. We have to find solutions where they happen to be, not where we would like them to be.

In the first instance, the new inputs from molecular and cell biology have their most obvious application in the identification of purified antigenic components. The possibility of dissecting the malarial plasmodium to find those antigenic constituents that can be used most effectively for vaccination is of course a very exciting challenge, and it is gratifying to see how rapidly William Trager's work in this field is being taken up in many laboratories that are investigating malaria.

That is the most obvious application and it was waiting in the wings for a technological breakthrough with which to exploit it. I hope this does not obscure a variety of other ways in which that type of experimental biological insight can be made indispensable. I call particular attention to the virtues of live vaccines. It should be possible to prepare them for any organism about which we have an intimate knowledge of its life cycle and control of its genetics.

When Louis Pasteur produced a rabies vaccine, without being aware of it he was selecting for less virulent strains because they have a lower neurotropism. We know very little today about the molecular basis of the altered strains or how the procedure used for cultivating them resulted in selection for less virulent mutants. Alfred Sabin was working equally in the dark when he developed live poliomyelitis vaccine. Even without knowing the fundamental biology one can proceed in a certain way if one has the technology to handle the organism.

But potential hazards lurk deep in both these situations precisely to the extent that we are not able to foresee what genetic interactions are still possible for those organisms. And of course the fact that we still need vaccinations for poliomyelitis means we have some residual problem with that disease. By that criterion the situation is far from perfect.

Perhaps even more important, the cell biology of the parasites could be considered in terms of understanding their development. They are, after all, eukaryotes that go through interesting and important developmental cycles. It is precisely those cycles that will show us targets of application for interventions that will drive a wedge between their developmental controls and metabolism and our own, because so much of what they do will arise from our common phylogenetic origins and the mutual adaptations of a parasite that has been living intimately in our bodies.

The fundamental perspective for any rational development of chemotherapeutic drugs merges very closely with that of pesticides, for the key to the control of both pests and parasites is an understanding of singularities of their developmental behavior compared to host and crop.

Another arena that tends to be put aside—and this is one of the tragedies of public health science in many areas where the most cogent intersection between the experimental laboratory and the public health perspective is needed—are "side effects." I cannot think of a more insidious expression for a set of phenomena that, far from being side effects, increasingly dominates our central policy concerns about the provision of agents to large populations.

Mention of the drug hycanthone gives some notion of what I am referring to. Many such controversies are impossible to resolve at the present stage of our knowledge of comparative toxology, of how to extrapolate data obtained under controlled laboratory conditions—and therefore perforce on a limited sample of individuals, ideally nonhumans—to what the risk will be when large populations are exposed.

Today, particularly, when questions of source, on the one hand, and target, on the other, of health research raise so many political problems, there is so much latent distrust that the potential magnitude of the so-called side effects dominates the situation more and more.

Side effects should very nearly occupy center stage in an examination of any program of practical intervention, rather than be left to the last minute as a kind of mop-up activity when the question is raised of whether untoward effects have been observed in an epidemiological survey, or when a laboratory animal is found to exhibit an unexpected side effect. The issue for physiologically potent chemicals is not whether they will have "side effects," but the details of potency, morbidity, and relationship of risks to benefits.

Any agent distributed to a great many individuals, under the inevitably poorly controlled conditions of a public health program, is bound to elicit some kind of imputed hazard. How can one be forearmed against this in ways that enable a rational examination of whether the danger is real or not? That is something that needs to be brought into our studies at the earliest stage. Simple empirical history tells us that, unfailingly, such accusations will be motivated and will be lodged against any agent introduced under those circumstances.

Finally, some reflections on medical progress in other fields that may bear some relationship to what we hope may result from this conference.

All of us who strive for a rational model of the universe we are exploring, and who work hard to improve the underpinnings, biological understanding, and chemical explanation of phenomena under study, believe that in the long run the use of basic science for health improvement is the most important route.

The problem is that things hardly ever work that way. If one examines the history of discoveries in the health field one finds no more than a minor sprinkling of circumstances where the framework of biological theory in any depth preceded our understanding of how to approach a particular disease.

The outstanding exceptions to that rule are the discoveries of Pasteur and Koch. They had their own prehistories of initial discoveries of bacterial agents of specific diseases. That territory having been opened up, in a few cases there was a crude theoretical framework for further exploration. But if we go back to 1830 instead of 1880, in terms of examining the history of that phenomenon, it falls within the paradigm I am about to express.

Almost every important advance in health practice, and in science generally, has emerged through Pasteur's paradigm, that is, "chance favors *the prepared mind*." (I would substitute "practical observation" for "chance.") That principle may be described as the natural history of what one sees in the test tube, at the bedside, in the field, or in public health observations of a disease: the confrontation with real problems; contradictions of one's theoretical expectations; insights about directions to take; essentially empirical discovery in the initial instance. Without immersion in a rather broad way in the complex system we are trying to deal with we simply will not have the encounters to provide the answers.

Those insights, those hopes, will be paralyzed at the outset, however, if they are not accompanied by the most sophisticated tools for further analysis. The discrepancies in our world view may not even be recognized unless a well-constructed theoretical framework exists to begin with. Their further exploitation will not be possible without many more basic scientific fundamentals than we are able to offer at a given time.

I have begun to despair of being able in my lifetime, or in those of two or three generations of our descendants, to fulfill the reductionist dream of a biological theory that will enable us to predict the nature of disease. That has been a misconception of the way fundamental science will eventually solve all our problems. I insist that we cannot make effective progress in the exploitation of any of the wide variety of empirical findings without that very sturdy base of further insights.

For this to work, there needs to be good communication among the different elements of the system. I do not expect every laboratory investigator to be able to spend a lot of time with researchers working in the field, or vice versa. But if people who are studying the natural history of disease are not in close communication with those who understand how to do the follow-up work we will miss the most important leads, and both health and science will suffer.

THE FUTURE OF PHARMACOLOGICAL RESEARCH IN PARASITOLOGY

Anthony Cerami

After considerable thought about the future of pharmacological research in parasitology I decided to discuss the problems inherent in this area because the development of drugs rests on the eventuality of being able to supply them to patients with parasitic diseases. The real problem is not how or where to find useful agents, but how to ensure that research to find new drugs continues.

Since the pioneering work of Paul Ehrlich on trypanosomes, which provided the basis for all of chemotherapy, there has been little activity in the area of producing drugs to prescribe for parasites. The virtual explosion of new antimicrobial and other pharmaceuticals introduced in the last twenty-five years leads us to question why there are so few safe and effective drugs for parasitic diseases. For example, no new trypanocidal agent has been discovered since 1940, and there are no reliable drugs for the treatment of Chagas' disease or onchocerciasis. The reason for this lies in part in the economics of the situation.

Drug development in the Western world in the last fifty years has evolved almost totally in the private sector, and the motivation has been, by necessity, profit. The increase in the cost of research to discover new drugs makes the selection of potential products a very careful process done under the guidance of many individuals who reflect the requirements of the corporation, which needs to find drugs that will eventually repay the investment it made to develop them. Before an agent can eventually be marketed the costs incurred can be significant, varying from several million dollars to as much as $20 million. Obviously this is a very risky business, for a drug may fail at any of a large number of places before it can be marketed.

Given these prerequisites one can understand why the pharmaceutical industry has been unwilling to make the necessary investment for the development of drugs for parasites that are not found in animals and humans in the Western world; it is financially too risky a venture to embark upon. That the industry has the capability and wherewithal necessary to produce new drugs is best evidenced by the significant number of agents available to treat coccidiosis, a disease caused by a parasite found in the United States and Europe. These drugs constitute a source of revenue of several hundred million dollars a year to the companies involved.

The statement that the drug industry is unwilling to take the risk necessary to produce new drugs is not meant as a criticism of the industry, but rather to point out the obvious fact that the private sector has to maintain a profit motive if it is to survive. It is obvious therefore that alternatives to the private sector need to be found. In this connection the Walter Reed Institute for Medical Research of the United States army has for many years been engaged in research on new agents for the treatment of malaria and other parasitic diseases of military import; the National Institute of Health, on the other hand, has not been involved to a significant degree.

It is apparent from the heated debates that have been going on for many years that the responsibility for the development of drugs for parasitic diseases is disputed; each party claims the other should be culpable. The drug industry, for example, points to the lack of infrastructure and stability in the Third World as a significant deterrent for the investment of funds, while Third World countries point to the drug industry as representing colonialist exploiters; moreover neither national nor international governmental agencies have been able to mount the kind of research programs necessary. As a result, over the past twenty-five years very little action has been taken and very few new drugs have been developed. Recent World Health Organization activities are encouraging the search for new agents, but such programs are by necessity of a limited nature, primarily small, investigator-initiated grants to scientists in the developing world.

In 1979, at a time when Senators Edward M. Kennedy and Jacob K. Javits were calling for an investigation of the industry to determine why nothing was being done to provide drugs to the less-developed countries, the National Academy of Sciences held a meeting on this issue. Representatives of industry, national and international governmental agencies, and academic institutions met to explore new approaches to the problem. A report of the meeting was subsequently published.[1] A more important outcome was the establishment of a task force to conduct a continuing dialogue among the various groups concerned; under the direction of James Henry of the Center for Public Resources, the task force has since been searching for solutions to the problem of drugs for the developing world. A number of meetings of the main task force and its subgroups have been held to explore new ways to improve the infrastructure of health care and drug distribution and utilization as well as new means for promoting the production of drugs for these countries.

As a result of these meetings the nonprofit Drug and Vaccine Development Corporation was created to help fill the void by trying to bring about a partnership of the academic community, industry, and government agencies. It is proposed that the research activities of selected groups be monitored in a search for possible patents that may lead to the development of new drugs and vaccines. The protection of these patents will be essential to ensure further progress because industry is reluctant to make the required investments without such protection. It is hoped that the procedure necessary to obtain a patent will be simplified and present no hindrance to publication of the information that is essential to scientists working in academic institutions. Every effort will be made to license the patents and make them available to industry. If further work is needed before a potential company can be located, then additional toxicological and pharmacological studies will be performed.

The model to be followed by the corporation is similar to that undertaken by the Population Council in the production and distribution of contraceptive agents. A board of directors has been appointed and it is anticipated that this fledgling corporation will be nurtured and allowed to fill some of the gaps that currently exist. This approach should lead to new applications of the advances in the development of new drugs and vaccines made in modern chemistry and biology over the last twenty-five years. It is hoped that over the next twenty-five years many of the agents for parasitic diseases will be produced if the necessary capital can be allocated for this important purpose.

NOTE

1. A. Cerami and S. R. Meshnick, "The Rational Development of New Drugs for Trypanosomiasis," in *Pharmaceuticals for the Developing Countries,* ed. J. H. Simon (Washington: National Academy of Sciences, 1979).

DISCUSSANT:

Seymour J. Klebanoff

I would like to discuss possible fresh applications of an old strategy for the development of new agents to control parasitic diseases.

When an antiparasitic agent is administered, selective toxicity to the parasite generally depends on the presence of a metabolic difference between parasite and host that can serve as the focus for attack. This is often a fine line, and toxicity to the host is the major limiting factor in the use of the agent. There is precedence, however, for the controlled use of the patient's own body as a source of toxic agents.

The body responds to foreign invasion with the production of extremely toxic agents, and it has devised means to deliver them in such a way that the invader is affected whereas the host is not. This is the normal immune response. It is possible, however, to manipulate or modulate the host response through external influences whereby the toxic agents are produced in greater than normal amounts; a physiological attack now becomes a pharmacological one, using the body's own sensing devices to direct the assault against the invading organism rather than against the host.

This concept is of course as old as immunology and the recognition that a specific antibody directed against the invader can be produced in greater than normal amounts by administration of the appropriate antigen. Attempts have been made in recent years to activate macrophages nonspecifically in humans by, for example, administering BCG in the hope that such cells will be more effective against foreign targets. This further demonstrates the strategy of directing the pharmacological agent against the host, with the foreign invader being affected by the host response. Are there other opportunities to use the patient's own body as a source of pharmacological agents and delivery systems?

Our interest has been in the production by phagocytes—neutrophils, eosinophils, macrophages—of a number of highly toxic agents effective against a variety of microorganisms. When the phagocyte cell membrane is perturbed—as by a ligand-receptor interaction with antibody and/or complement-coated particles, or by exposure to a variety of soluble stimulants—there is a burst of oxygen consumption, and much if not all the extra oxygen consumed is converted to the one electron reduction product, the superoxide anion (O_2^-), which is not itself highly toxic. It is capable of certain oxidations and reductions, but these reactions do not appear to be sufficiently harmful to cause the organisms to die. Rather, the superoxide anion appears to exert its toxicity through the products it forms (Figure 1).

Hydrogen peroxide (H_2O_2) is generated by the dismutation of the superoxide anion, a reaction in which two molecules of that anion interact, with one being reduced and the other oxidized with the formation of oxygen and hydrogen peroxide as follows:

$$O_2^- + O_2^- + 2H^+ \rightarrow O_2 + H_2O_2.$$

The highly reactive hydroxyl radical (OH·) can be generated by a metal-catalyzed interaction between superoxide and H_2O_2, the so-called Haber-Weiss reaction,

$$O_2^- + H_2O_2 \rightarrow O_2 + OH^- + OH\cdot,$$

and this same reaction may generate oxygen initially in its excited form, that is, as singlet molecular oxygen ($^1O_2$). All these products of the respiratory burst of phagocytes, that is, H_2O_2, OH·, and possibly $^1O_2$, have potent toxic activity, with the toxicity of H_2O_2 being greatly potentiated by the addition of peroxidase and a halide.

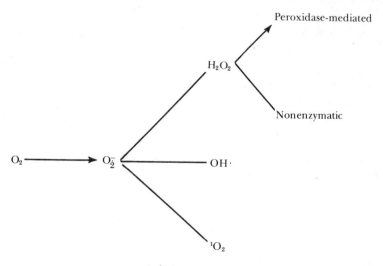

Figure 1. Production of toxic oxygen products.

The role of peroxidases in the cytotoxic activity of phagocytes is an area in which we are particularly interested. A
peroxidase—myeloperoxidase (MPO)—is present in exceptionally high concentration in cytoplasmic granules of neutrophils,
and the same enzyme is found in monocytes. A different
peroxidase—eosinophil peroxidase (EPO)—is present in
eosinophil cytoplasmic granules. When phagocytes are stimulated, as by an appropriately opsonized parasite, the respiratory
burst is accompanied by the release of peroxidase—as well as
the other components of the cytoplasmic granules—either into
the phagosome when the parasite is ingested, or extracellularly
when phagocytes adhere to organisms too large to be ingested.
The peroxidase reacts with H_2O_2—formed either by leukocytic
or, for some organisms, microbial metabolism—and a halide
(chloride, bromide, iodide) to form a very powerful toxic system. The peroxidase and H_2O_2 form an enzyme-substrate
complex that oxidizes the halide to form an agent or agents
with potent toxic properties. With chloride as the halide, the
primary oxidant formed is hypochlorous acid, a powerful ger-
micidal agent probably responsible for much of the toxicity of
the peroxidase system in neutrophils. Other agents also are
generated from this and other halides, and these very reactive
species attack the target cell with both halogenation and oxida-
tion of surface components.

The eosinophil peroxidase has an additional property of
interest in relation to its role in the host defense against certain
parasites. It is a strongly basic protein that adheres to negatively
charged surfaces, including the surface of parasites. The organism with peroxidase on its surface becomes highly sensitive
to H_2O_2, which can be generated by any phagocyte in the
region. Thus, for example, macrophages that do not contain a
peroxidase in cytoplasmic granules, but do generate H_2O_2, can
utilize that H_2O_2 more efficiently when the organism under
attack, either in a phagosome or extracellularly, is coated with
peroxidase. This is an example of the synergism of phagocytic
cells in inflammatory foci; one cell provides the peroxidase and
another the H_2O_2, with the total effect greater than the sum of
its parts.

The agents released by phagocytes are not limited to the

toxic products of oxygen metabolism. Neutrophils and, in particular, eosinophils also contain strongly basic proteins in cytoplasmic granules that are toxic nonenzymatically. The hydrolytic enzymes released can attack a variety of macromolecules, and there is recent interest in the formation of arachidonic acid metabolites by stimulated phagocytes that are tremendously powerful biological mediators.

Thus phagocytes are rather dormant metabolically when at rest, but when stimulated they produce and release some of the most reactive agents known to man. Further, they are attracted to invading organisms, including parasites, and may adhere to these cells and thus be in a position to release toxic components in close proximity to the target. Can we manipulate these systems by the administration of pharmacological agents so the attack becomes more effective? The administration of agents that stimulate peroxidases, generate H_2O_2 or oxygen radicals, or increase the available halide pool are possible approaches. The goal of this strategy is to utilize more efficiently both the very powerful toxic agents produced by the body and the delivery systems that direct the agents against an invader while sparing the host.

A NEW LOOK AT THE ELEPHANT

Irwin W. Sherman

The roles played by biochemistry and pharmacology in studies of parasitism are evident, for the present degree of control of disease-producing organisms is largely due to the effective use of chemical agents that eliminate invading pathogens and prevent vectors from successfully transmitting these agents. If pharmacology and biochemistry provide such a powerful chemical arsenal, why then have the problems of parasitic diseases not been solved?

In part this is because the parasitologist, pharmacologist, and biochemist have not effectively combined their special talents and training to deal with the problems at hand. They view the host-parasite relationship as the blind men did the elephant (Figure 1).

Undoubtedly you recall the poem by John Godfrey Saxe, which reads, in part:

> It was six men of Indostan
>   To learning much inclined,
> Who went to see the Elephant
>   (Though all of them were blind),
> That each by observation
>   Might satisfy his mind.
>   . . . .
> The *Third* approached the animal,
>   And happening to take
> The squirming trunk within his hands,
>   Thus boldly up and spake:
> "I see," quoth he, "the Elephant
>   Is very like a snake!"

The *Fourth* reached out an eager hand,
  And felt about the knee.
"What most this wondrous beast is like
  Is mighty plain," quoth he;
" 'Tis clear enough the Elephant
  Is very like a tree!"
. . . .
The *Sixth* no sooner had begun
  About the beast to grope,
Than, seizing on the swinging tail
  That fell within his scope,
"I see," quoth he, "the Elephant
  Is very like a rope!"

And so these men of Indostan
  Disputed loud and long,
Each in his own opinion
  Exceeding stiff and strong,
Though each was partly in the right,
  And all were in the wrong!

MORAL:

So oft in theologic wars,
  The disputants, I ween,
Rail on in utter ignorance
  Of what each other mean,

And prate about an Elephant
Not one of them has seen!

John Godfrey Saxe
"The Blind Men and the Elephant"

Even a casual glance at the literature on parasites demonstrates the point Saxe made poetically. Over the decades research interest has waxed and waned, but even during periods of intense activity a fragmented approach toward unraveling the mechanisms that underlie the host-parasite interaction has persisted. The complexity of the host-parasite ecosystem requires collaborative work on the part of investigators in different disciplines, but it is the exception rather than the rule for a dialogue to be established among the biochemist, pharmacologist, and parasitologist. It is also rare to find individuals well trained to work effectively in all three areas.

This lack of communication stems in part from the traditional structure of the curriculum in those institutions where parasitology is biology, biochemistry is chemistry, and pharmacology is medicine. Because training in parasitology falls into the category of "zoology," in-depth coverage of the "physical and chemical" tends to be neglected. In those institutions where the teaching of parasitology involves mechanistic approaches, the course content is often not at the cutting edge. In some cases this deficiency derives from a lack of time for "biochemical" electives, in others it simply reflects the rapid acceleration in the mass of knowledge to be acquired in biochemistry and pharmacology and the inertia inherent in movements to bring about curricular reform. For future training the removal of the blinders will require curricular changes with greater emphasis on the physical sciences. Indeed a new definition of parasitism will necessitate investigations based on the combined efforts of individuals who are prepared to share their special skills so that the final synthesis is one that could not have been attained by a single specialized approach.

Curricular reform must also entail changes in the approach and attitude of the experimenter. Research in biochemistry and pharmacology tends to center around functional questions; particular organisms admirably suited for the experimental approach are selected later. Parasitologists, on the other hand, tend to work in reverse: the disease organism is chosen first, and only then are functional questions asked. Thus the biochemist working on muscle contraction may use rabbit or scallop muscles as a source of experimental material because of their abundance and the ease with which they may be isolated. The parasitologist studying blood fluke locomotion is restricted to microscopic worms that are often difficult to obtain in the amounts necessary for biochemical analysis.

A partial solution is to make parasites readily available in the quantities required for current biochemical methodologies by "farming" parasites in vivo or in vitro, or, alternatively, scaling down the analytic methodologies. If parasites are provided in abundance as shelf items, and if training programs assure that future workers are capable of dealing with the molecular mechanisms that underlie parasitism, there is every reason to believe that more effective "magic bullets" will be designed to fell the parasitic intruders.

What are the unique attributes of parasites that might be exploited by the biochemist and pharmacologist for drug design and production? The examples I have chosen involve malarial parasites simply because they are the organisms with which I am most familiar.

One principle of drug design is that the agent or an active metabolite of it be accumulated by the target organism so that selective toxicity may occur. In an intracellular parasite such as malaria the drug would first contact the interface of the plasma membrane of the host cell. An understanding of the architecture and composition of the host cell's plasma membrane would therefore contribute to the design of more effective antimalarials. Is it not possible to describe membrane receptors for drugs? And are there not membrane properties of the infected host cell that might be utilized to enhance drug accumulation? Information derived primarily from animal models demonstrates that both the lipid and protein composition of the malaria-infected cell is altered (Tables 1–3). Could such information not be deployed for the delivery of drugs by liposomes?

One spectacular advance of the past twenty-five years has

TABLE 1. LIPID CONTENTS OF ERYTHROCYTES AND PLASMODIUM

	Ratio of Phospholipid to Cholesterol
Monkey	
Red blood cell	2.15
P. knowlesi	6.3
Duck	
Red blood cell	3.6
P. lophurae	10.7
Rat	
Red blood cell	2.2
P. berghei	?
Human	
Red blood cell	2.7
P. falciparum	?

been the deciphering of the genetic code, which has clearly shown that species differ in their nucleic acid composition. Because malarial parasites show unique DNA and ribosomal RNA (rRNA) base compositions (Table 4), there is good reason to be optimistic that drugs can be tailored to selectively inhibit the synthesis and function of plasmodial nucleic acids without harming the host.

Related to the uniqueness in plasmodial nucleic acids is the

TABLE 2. DISTRIBUTION OF SOME SATURATED FATTY ACIDS IN TOTAL LIPIDS

	Percent Distribution			
	16:0	18:0	18:1	Total
Monkey				
Red blood cell	31	16	16	63
P. knowlesi	33	11	35	79
Rat				
Red blood cell	24	17	8	49
P. berghei	42	15	21	78
P. vinckei	40	13	17	70
Duck				
Red blood cell	24	10	18	52
P. lophurae	26	16	33	75
Human				
Red blood cell	27	9.4	19.5	56
P. falciparum	?	?	?	?

TABLE 3. MEMBRANE PROTEIN ALTERATIONS IN MALARIA-INFECTED RED CELLS

	Bands Decreased	Bands Increased	New Bands
P. berghei	1 (220k)		
	2 (210k)		2a (140–165k)
	3 (118k)		
	4 (108k)		
			5 (42k, phosphorylated)
		7 (54k)	
		8 (44k)	
			9 (20k)
P. chabaudi	1		
	2		(150k)
	4		
P. vinckei			(150k)
P. yoelii YM			
17X	1, 2		(120–210k) weak
33X			(120–210k) weak
P. knowlesi	1 (300k)		
	2 (240k)		2a (120–220k)
	5 (52k)		
		4a (68k)	
P. falciparum	1 (240k)		
	2 (215k)		
	3 (88k)		
P. lophurae	1 (230k)		
	2 (210k)		
	3 (88k)		
		4 (53k)	
		5 (43k)	
		6 (32k)	
		7 (20k)	

NOTE: Numbering of the bands is in order of decreasing molecular weight; molecular weight in daltons is given in parentheses.

fact that all the proteins of the malarial parasite must differ from those of the host. Several years ago we described unique lactic and malic dehydrogenases from a bird malaria, each with different catalytic properties than those of the host cell (Tables 5 and 6).[1,2] More recently other isoenzymes from mammalian and human species of malaria have been described,[3] but to my knowledge little use has been made of this data in drug design.

Just over a decade ago it was discovered that malarial

TABLE 4. BASE COMPOSITION OF MALARIAL PARASITES AND THEIR HOSTS

	Percent G + C	
	DNA	Ribosomal RNA
P. lophurae	20,19	35
P. gallinaceum	18	—
Duck reticulocyte	39	64
P. knowlesi	37,19	37
Money reticulocyte	—	67
P. falciparum	37,19	—
P. berghei	24	37–43
P. vinckei	24	—
P. chabaudi	27,17	—
Mouse liver	40	58–61

parasites were incapable of de novo purine biosynthesis and thus were dependent on salvage mechanisms. In line with this was the finding that, although the parasites lacked the folate enzymes concerned with the pathway for purine biosynthesis, those involved with the de novo pyrimidine pathway were present.[4] It was also found that the parasite enzyme involved in the formation of tetrahydrofolate—dihydrofolate reductase— had a higher molecular weight than the host enzyme and a binding constant for pyrimethamine (Table 7) several orders of magnitude greater than the host enzyme.[5] These unique attributes undoubtedly contribute to the exquisite sensitivity of the parasite to pyrimethamine, but the molecular basis of sensitivity and how drug resistance develops still remain unclear. Clearly this is fertile ground for future study.

TABLE 5. RATIOS OF REACTION RATES OF DUCK ERYTHROCYTES AND
Plasmodium lophurae LACTIC DEHYDROGENASE WITH REDUCED
NICOTINAMIDE-ADENINE DINUCLEOTIDE (NADH) ANALOGS AND PYRUVATE

	Duck Red Blood Cell	P. lophurae
DeNADH (1) / NADH (3)	1.97	0.455
NADH (3) / NADH (1)	0.47	0.94

NOTES: (1) = 3×10^{-4} M pyruvate. (3) = 1×10^{-2} M pyruvate. DeNADH = deamino NADH.

TABLE 6. RATIOS OF REACTION RATES OF DUCK ERYTHROCYTES AND
Plasmodium lophurae MALIC DEHYDROGENASE WITH ANALOGS OF NAD

	Duck Red Blood Cell	*P. lophurae*
APNAD (3) / APNAD (1)	0.36	0.77
DeNAD (3) / DeNAD (1)	0.91	1.60
APNAD (1) / NAD (1)	9.5	0.8
NAD (3) / TNNAD (3)	2.3	1.1

NOTES: (1) = 6×10^{-3} M malate. (3) = 1×10^{-1} M malate. APNAD = 3-acetylpyridine NAD. DeNAD = deamino NAD. TNNAD = thionicotinamide NAD.

Based on these examples we might ask: If the metabolism of the parasite differs to such a large degree from that of the host, why are there so few examples of drugs tailored to take advantage of this uniqueness?

In part our limitations relate to the way we look at the "Elephant." Parasite biochemistry takes its cues from biochemical studies of prokaryotes, free-living eukaryotes, and the tissues of eukaryotes. During the early years of biochemistry emphasis was placed on establishing its unity. Thus much of the classical research was concerned with the commonality of energy production in organisms and the universality of genetic

TABLE 7. SENSITIVITY OF DIHYDROFOLATE REDUCTASE TO PYRIMETHAMINE

	Pyrimethamine Concentration ($\times 10^{-8}$ M) for 50 percent Inhibition
P. knowlesi	0.1
Human red blood cell	180
P. berghei	.05–0.1
Resistant strain	2.0–3.0
P. vinckei	0.2
Resistant strain	45
Mouse red blood cell	100
P. lophurae	0.6
Duck erythrocyte	19

mechanisms. By analogy this is comparable to an individual visiting the basements of the National Gallery, the White House, and the Library of Congress, finding a furnace in each, and then, without going upstairs, concluding that these buildings were identical. It is only relatively recently that biochemists, pharmacologists, and parasitologists have begun to look around "upstairs" where the uniqueness lies. This new focus will certainly benefit future drug design.

Once the uniqueness of enzymes is fully explored we will be in a much better position to design drugs that will selectively bind to and inhibit some essential parasite reaction without affecting the host. Most intriguing in this regard is the development of "suicide substrates"—unreactive synthetic substrates, which, at the active site of an enzyme, are converted into a reactive species, which in turn react specifically with the enzyme to inactivate it. The nature of the enzyme active site-substrate interaction itself provides for built-in selectivity.

It is evident that the design of chemotherapeutic agents so desperately needed for the control of malaria and other parasitic diseases requires multidisciplinary approaches. We will need a revision of the academic curriculum; we have to develop means for retraining and refocusing the interest of biochemists and pharmacologists so parasites become favored objects of study; and we have to provide parasites in sufficient amounts and of such quality that they can be used effectively by those asking mechanistic questions. By a concerted approach, and in the not-too-distant future, we should be able to describe the parasite's molecular fingerprint, a unique mark that will determine the means for its eradication.

At that time we will look upon the "Elephant" in another way (Figure 2):

> With their blinders off, and confusion gone,
>   "It's an Elephant," they conceded,
> Each one had a clearer perception
>   Of what the problem needed.
> Characteristics known, no longer was there fear.
>   "O' million murdering deaths," the end it seems is near.*

---

* Last stanza by the author, with an acknowledgment to Sir Ronald Ross for the phrase, "O' million murdering deaths."

NOTES

1. I. W. Sherman, "Heterogeneity of Lactic Dehydrogenase in Avian Malaria Demonstrated by the Use of Coenzyme Analogs," *Proceedings of the First International Congress on Parasitology* 1 (1964): 73.

2. ————, "Malic Dehydrogenase Heterogeneity in Malaria (*Plasmodium lophurae* and *P. berghei*)," *Journal of Protozoology* 13 (1966): 344–49.

3. ————, "Biochemistry of *Plasmodium* (Malarial Parasites), *Microbiology Reviews* 43 (1979): 453–95.

4. E. Platzer, "Metabolism of Tetrahydrofolate in *Plasmodium lophurae* and Duckling Erythrocytes," *Transactions of the New York Academy of Sciences*, ser. 2, 34 (1972): 200–08.

5. R. Ferone, "Folate Metabolism in Malaria," *Bulletin of the World Health Organization* 55 (1977): 291–98.

THE ROLE OF IMMUNOLOGY IN STUDIES OF PARASITIC INFECTIONS WITH SPECIAL REFERENCE TO SCHISTOSOMIASIS

Adel A. F. Mahmoud

INTRODUCTION

Immunology, a relatively old scientific discipline, is intimately related to infectious diseases. The subject matter of immunology has expanded over the past few decades to encompass a wide range of study areas such as host responses to pathogens, immunopathogenetic mechanisms of disease, tumor immunology, and tissue transplantation.

Immunology attempts to provide mechanistic explanations for the interaction of self with nonself. In fact its beginning was based on studies of infectious disease models. The application of immunological tools to the study of parasitic infections in their current narrow definition, that is, pathogens of animal origin, started almost simultaneously with studies of the immunology of other infectious agents such as bacteria and viruses. For reasons that remain unclear, however, the progress of immunological studies in the field of parasitology has been slow. Serodiagnosis and antibody responses were the starting point, but, while microbiologists and immunologists moved to explore and expand their knowledge of the host-parasite relationship in bacterial and viral infections, immunological studies of animal parasites lagged.

184

The last two decades have witnessed an aggressive attempt by parasitologists, immunologists, and others to apply immunological knowledge to the study of parasites and their hosts. Immunology did not and does not promise to provide answers to all these problems; rather, it offers a variety of approaches that will lead to a better understanding of the host-parasite relationship. Such an understanding is crucial if we are ever to succeed in manipulating this relationship to the advantage of the host.

It is impossible in this chapter to provide an overview of what immunology can offer to our common area of interest, the study of human and veterinary parasites. I will therefore concentrate on the role it has played in the laboratory and in the field in studies of one parasitic infection: schistosomiasis. There are many reasons for such a biased choice, not the least being the fact that immunology contributed significantly—probably more than to any other parasitic infection—to our current understanding of schistosomiasis and its disease sequelae.

SCHISTOSOMIASIS: HISTORICAL BACKGROUND

Hematuria is known to have occurred among ancient Egyptians, who named it the "aaa" syndrome and found that it came from the waters of the Nile. To avoid infection, people were advised to cover their external genitalia when going into the river. Almost 4,000 years later, in A.D. 1851, a young German pathologist, Theodore Bilharz, discovered the fascinating digenic trematode. The decades to follow were spent in trying to determine the life cycle of the parasite and its clinical and pathological consequences. *Schistosoma haematobium* causes bladder and kidney lesions; in some individuals *S. mansoni* infection causes hepatosplenomegaly associated with portal hypertension and esophageal varices. Pathological and clinical studies characterized the liver lesions and the types of portal fibrosis and defined the hemodynamic changes.

In contrast, the contributions of immunological studies were limited to the development of serological methods for the diagnosis of infection using crude antigen preparations. Preliminary attempts were initiated to study immunity in schistosomiasis. Immunology therefore contributed very little to studies of schistosomiasis during the first century following discovery of the worms.

In contrast, the past thirty years have witnessed an exponential explosion of research on the immunology of schistosomiasis. Besides providing new investigative and diagnostic tools, these studies have resulted in major discoveries of the pathogenesis of disease and immunity in this infection. Two examples will be discussed in detail: the development of our current understanding of granuloma formation and pathogenesis of disease and its modulation; and the progress of our knowledge of immunity in schistosomiasis.

PATHOGENESIS OF DISEASE IN
SCHISTOSOMIASIS MANSONI

The development of our understanding of the schistosome egg granuloma and its principal role in disease etiology is summarized in Table 1. With the progress made in immunological studies by the late 1950s, the stage was set to apply these and other techniques to investigate schistosomal disease. Although it had long been known that the parasite eggs retained in the host tissues are surrounded by massive inflammatory lesions, the significance of the observation was not clear. In the early 1960s F. von Lichtenberg introduced the lung granuloma model, in which isolated *S. mansoni* eggs are injected into the pulmonary microvasculature of mice via a tail vein; by sectioning the lungs at time intervals thereafter a quantitative measure of granuloma formation can be obtained.[1]

K. S. Warren and his colleagues played the central role in establishing the nature and mechanism of granuloma formation in schistosomiasis. They posed the right kinds of questions: Is the lesion immunological in nature? What is its specificity and its mechanism? In a series of studies they established that granuloma formation in *S. mansoni* is an immunological phenomenon.[2-4] Animals not previously exposed to the parasite or to its eggs form granulomas of a certain size that peak in sixteen days. Prior exposure of mice to eggs results in the formation of larger granulomas that peak in eight days. Sensiti-

TABLE 1. CONTRIBUTION OF IMMUNOLOGICAL INVESTIGATIONS
TO THE UNDERSTANDING OF THE SCHISTOSOME EGG GRANULOMA
AND ITS ROLE IN DISEASE IN EXPERIMENTAL ANIMALS

Observation	Significance	Notes[1]
Lung granuloma model	Introduction of quantitative techniques in studies of schistosomiasis pathology	1
Immunological nature of egg granuloma	Recognition of the mechanism of the basic pathological lesion	2
Etiology of egg granuloma	• Cell-mediated response • Role of T lymphocytes and parasite antigens in pathogenesis	3–6
Granuloma culture in vitro	Role of lymphokines and antibodies in induction and regulation of disease	7,8
Granulomas and liver fibrosis	Types and rates of collagen synthesis and deposition in schistosomiasis	9,10
Modulation of granuloma formation in vitro and in vivo	Role of lymphocyte subsets and mechanisms of modulation	11–15
Role of eosinophils in granulomas • in vitro • in vivo	 Eosinophils destroy eggs Eggs are retained in tissues of eosinophil-depleted animals	 16–18 19,20

NOTE: 1) Refers to notes at end of this chapter.

zation is achieved by schistosome eggs, not by any other parasites; that is, it is specific. Finally, the mechanism of granuloma formation is described as cell-mediated because it is transferred by lymphoid cells, not by serum. Confirmation of this concept was achieved by many studies, including the use of antilymphocyte serum, neonatal thymectomy, bursectomy in chicken, and anti-μ serum in mice. The responsible antigens and cellular composition of granulomas have also been described.[5,6]

Thus the stage was set for a better understanding of *S. mansoni* egg granuloma, its physiological and pathological role in human disease, and its relationship to the permanent fibrotic lesions.[7–10] Fascination with the phenomenon of granuloma formation continues; mice with chronic *S. mansoni* develop smaller lesions and the consequences of egg granulomas are milder. This amelioration, originally described as "endogenous desensitization," is now termed "modulation." It involves the development of specific subsets of suppressor T cells, which results in a decrease of granuloma size in chronically infected animals.[11–15]

Immunological studies have also contributed to our understanding of the functional role of the host granulomatous response. Although it was suspected that granulomas lead to egg destruction, there was no formal proof for such a role. S. L. James and D. G. Colley were the first to introduce in vitro techniques to examine the interaction of S. mansoni eggs and host cells.[16–18] Their elegant studies specifically examined the possible role of eosinophils, antibodies, and lymphokines in the destruction of eggs, and, in the process, they defined a functional role for the eosinophils. More recently we examined the fate of eggs in host tissues in vivo, calculated destruction rates, and examined the role of the different cellular components within these granulomas.[19,20] Eosinophils were shown to be the primary cells causing egg destruction in vivo; in their absence, eggs accumulate in tissues and, in infected mice, fetal disease develops.

Simultaneous with these experimental studies, immunological tools were introduced to understand man's response to schistosome worms, to explain some of the observed phenomena in human disease, and to maintain the search for approaches to control the infection or its disease sequelae.[21] For generations scientists engaged in research on schistosomiasis and other helminth infections have recognized the heterogeneity of parasite distribution in endemic communities. Epidemiological and parasitological surveys have shown that most infected individuals harbor a few worms, and that only a small proportion is heavily infected. Furthermore, disease, as manifested by hepatosplenomegaly, is again seen in no more than 10 to 15 percent of infected individuals.

This curious distribution of infection and disease has led to speculation that it is due to the development of immunity, but no proof has yet been offered. Alternatively, the pattern of contact of individuals in endemic areas with bodies of infected water may explain this heterogeneity. A third explanation—particularly in cases where all experimental evidence points to a central role for the host immune response—is that heterogeneity may be due to variations in man's response to the schistosome infection. This hypothesis, if valid, should revolutionize our understanding of schistosomiasis in humans and force us to pause and consider how control programs should be designed.

In a series of studies performed in the field, we have found that schistosome-infected individuals react in a heterogeneous way to parasite-derived antigens.[22] For example, stimulation of lymphocyte DNA synthesis was found to be inversely proportional to the intensity of the infection. The defect in heavily infected individuals is specific; it is not seen when other antigens or nonspecific mitogens are used. Similarly, heavily fected individuals have a severe defect of myocyte killing of schistosomula.

These observations strongly suggest, therefore, that the heterogeneity in schistosomiasis distribution in endemic communities may be due to a spectrum of host immunological responses, and that there might be a subset of individuals susceptible to infection and disease. Moreover, we have shown that the development of schistosomal hepatosplenomegaly may be genetically based. Individuals who happen to carry the HLA-A_1 and B_5 haplotypes are at particular risk of developing disease following infection.[23] In fact the susceptibility is so strong that when both haplotypes are present all individuals examined have moderate to severe hepatosplenomegaly.

To summarize, immunological tools have provided scientists working on schistosomiasis with an opportunity to examine the pathogenesis of disease at the cellular and molecular levels, to conduct studies to better understand disease in man, and, one hopes, to explore methods for manipulations that will benefit the host.

PROTECTIVE RESPONSES IN SCHISTOSOMIASIS

The other area I will discuss concerns the host's protective mechanisms against schistosomiasis. The schistosome worms, like other multicellular parasites, have developed a multitude of systems to avoid the host immune response. The host, however, continues to utilize several defense mechanisms, operating at many levels to limit the infection or its disease sequelae. A few years ago we knew very little of the host protective mechanisms in schistosomiasis. But, due to the interest of modern immunologists in studies of parasitic infections, we now know that in vitro and in vivo there are several mechanisms that can kill the invading forms of the schistosomes.

The progress of our knowledge of resistance and its mechanisms in schistosomiasis paralleled the introduction of modern immunological concepts and tools. It has been known for years that some animal species develop resistance to subsequent challenge with schistosome worms.[24,25] Furthermore, species-dependent variability in host susceptibility to infection has been recognized.[26] The development of our understanding of resistance to schistosomiasis is outlined in Table 2.

Following the demonstration of acquired resistance, two key studies paved the way to the current status of our knowledge: 1) the passive transfer of immunity by serum in certain animal species;[27] and 2) the development of in vitro models to study host-parasite interactions.[28] The role of the eosinophil in host protection and the mediators of parasite destruction have been examined extensively.[29–33] The eosinophil seems to play a central role in host defense. It was difficult, however, to imagine how a small cell, with a diameter of approximately 14 μm, could attack and kill a multicellular parasite of more than 200 μm in length. The biochemical and subcellular events that lead to the death of the parasite are described by Seymour J. Klebanoff in his chapter in this book. In vitro, the interaction of eosinophils and parasites go through stages of attachment, deposition of granule content, penetration of the helminth surface, and finally its disintegration.[34] The specific deadly message delivered by the eosinophil during this interaction may be the contents of its granule, either the major basic protein or products of peroxidative metabolism.[35–40]

Other protective mechanisms have recently been identified; of note is the role of macrophages and monocytes in what has been termed "nonspecific" acquired resistance.[41,42] This cell line may not only be involved in acquired mechanisms of resistance, but its activity may reflect innate protective mechanisms.

TABLE 2. CONTRIBUTION OF IMMUNOLOGICAL INVESTIGATIONS TO THE UNDERSTANDING OF RESISTANCE TO SCHISTOSOMIASIS IN EXPERIMENTAL ANIMALS

Observation	Significance	Notes[1]
Animals with primary infection are partially protected	Acquired protection may exist in schistosomiasis; resistance is species-related	21,24,25
Serum from animals with chronic infection partially protects recipients	Antibody-dependent mechanism for protection	26–28
Antieosinophil serum blocks acquired immunity	Resistance is antibody-dependent and is mediated by eosinophils	29,30
Development of in vitro models for host interaction with parasite	Role of antibody, eosinophils, neutrophils, complement, and host macrophages in resistance to schistosomiasis	31–35
Mechanisms of cell-mediated cytotoxicity	Role of peroxidative pathway	36,37
	Eosinophil and neutrophil basic proteins	38,39
	Arginase	40
Induction of resistance by "nonspecific" agents	Activated macrophages play a crucial role in host resistance	41,42
Induction of resistance by irradiated parasites	Protection of experimental animals and cattle in the field is possible	43–45

NOTE: 1) Refers to notes at the end of this chapter.

WHAT CAN IMMUNOLOGY OFFER?

I will close this presentation by addressing a crucial question: "What can immunology offer parasitology?" There is no simple answer to this question except to state that much has already been gained by applying the tools of immunology to study parasitic infections; schistosomiasis is a very good example. Immunology is one of the fastest developing scientific disciplines. It will continue to provide new tools and new knowledge that will have to be applied in order to better understand the host-parasite relationship in any given infection. It has to be made clear, however, that immunological tools are not going to provide vaccines or miracles for all human infections.

To be specific, all scientific disciplines, including immunology, provide the base of information that is critical if we are ever to succeed in our attempts to induce protection against parasitic infections.[41–43] The anti-intellectual and counter-productive arguments about the necessity for applied research and the secondary role of basic investigations will lead to another period of stagnation. On the other hand, oversimplification and raising false hopes will be just as counterproductive.

NOTES

1. F. von Lichtenberg, "Host Response to Eggs of S. mansoni. I. Granuloma Formations in the Unsensitized Laboratory Mouse," American Journal of Pathology 41 (1962): 711–31.

2. K.S. Warren, E.O. Domingo, and R.B.T. Cowan, "Granuloma Formation Around Schistosoma mansoni Eggs as a Manifestation of Delayed Hypersensitivity," American Journal of Pathology 51 (1967): 735–56.

3. K.S. Warren, "The Immunopathogenesis of Schistosomiasis: A Multidisciplinary Approach," Transactions of the Royal Society of Tropical Medicine and Hygiene 66 (1972): 417–34.

4. K.S. Warren, D.L. Boros, L.M. Hang, et al., "The Schistosoma japonicum Egg Granuloma," American Journal of Pathology 80 (1975): 279–94.

5. D.L. Boros and K.S. Warren, "Delayed Hypersensitivity-Type Granuloma Formation and Dermal Reaction Induced and Elicited by a Soluble Factor Isolated from Schistosoma mansoni Eggs," Journal of Experimental Medicine 132 (1970): 488–507.

6. D.L. Moore, D.I. Grove, and K.S. Warren, "The Schistosoma mansoni Egg Granuloma: Quantitation of Cell Populations," Journal of Pathology 121 (1977): 41–50.

7. D.L. Boros, R.P. Pelley, and K.S. Warren, "Spontaneous Modulation of Granulomatous Hypersensitivity in Schistosomiasis mansoni," Journal of Immunology 114 (1975): 1437–41.

8. K.S. Garb, A.B. Stavitsky, and A.A.F. Mahmoud, "Dynamics of Antigen and Mitogen Induced Responses in Murine Schistosomiasis japonica: In Vitro Comparison Between Hepatic Granulomas and Splenic Cells," Journal of Immunology (1981), in press.

9. M.A. Dunn, M. Rojkind, K.S. Warren, et al., "Liver Collagen Synthesis in Murine Schistosomiasis," Journal of Clinical Investigation 59 (1977): 666–74.

10. D.J. Wyler, S.M. Wahl, and L.M. Wahl, "Hepatic Fibrosis in Schistosomiasis: Egg Granulomas Secrete Fibroblast-Stimulating Factor in Vitro," Science 202 (1978): 438–40.

11. D.G. Colley, "Adoptive Suppression of Granuloma Formulation," Journal of Experimental Medicine 143 (1976): 696–700.

12. E.O. Domingo and K.S. Warren, "Pathology and Pathophysiology of the Small Intestine in Murine Schistosomiasis mansoni, Including a Review of the Literature," Gastroenterology 56 (1969): 231–40.

13. S.W. Chensue and D.L. Boros, "Modulation of Granulomatous Hypersensitivity. I. Characterization of T Lymphocytes Involved in the Adoptive Suppression of Granuloma Formation in Schistosoma mansoni-Infected Mice," Journal of Immunology 123 (1979): 1409–14.

14. S.W. Chensue, D.L. Boros, and C.S. David, "Regulation of Granulomatous Inflammation in Murine Schistosomiasis. In Vitro Characterization of T Lymphocyte Subsets Involved in the Production and Suppression of Migration Inhibition Factor," Journal of Experimental Medicine 151 (1980): 1398–1412.

15. D.G Colley, F.A. Lewis, and C.W. Todd, "Adoptive Suppression of Granuloma Formation by T Lymphocytes and by Lymphoid Cells Sensitive to Cyclophosphamide," *Cellular Immunology* 46 (1974): 192–200.

16. S.L. James and D.G. Colley, "Eosinophil-Mediated Destruction of *Schistosoma mansoni* Eggs," *Journal of the Reticuloendothelial Society* 20 (1978): 359–74.

17. ———, "Eosinophil-Mediated Destruction of *Schistosoma mansoni* Eggs in Vitro. II. The Role of Cytophilic Antibody," *Cellular Immunology* 38 (1978): 35–47.

18. ———, "Eosinophil-Mediated Destruction of *Schistosoma mansoni* Eggs. III. Lymphokine Involvement in the Induction of Eosinophil Functional Abilities," *Cellular Immunology* 38 (1978): 48–58.

19. A.A.F. Mahmoud, K.S. Warren, and R.C. Graham, Jr., "Antieosinophil Serum and the Kinetics of Eosinophilia in *Schistosomiasis mansoni*," *Journal of Experimental Medicine* 142 (1975): 560–74.

20. G.R. Olds and A.A.F. Mahmoud, "The Role of Host Granulomatous Response in Murine *Schistosomiasis mansoni*: Eosinophil-Mediated Destruction of Eggs," *Journal of Clinical Investigation* 66 (1980): 1191–99.

21. S.R. Smithers and R.S. Terry, "The Immunology of Schistosomiasis," *Advances in Parasitology* 14 (1976): 399–422.

22. J.J. Ellner, G. Richard Olds, G. Osman, et al., "Dichotomies in the Reactivity of Worm Antigen in Human *Schistosomiasis mansoni*," *Journal of Immunology*, 126 (1981): 309–12.

23. E. Abdel Salam, S. Ishaac, and A.A.F. Mahmoud, "Histocompatability-Linked Susceptibility for Hepatosplenomegaly in Human *Schistosomiasis mansoni*," *Journal of Immunology* 123 (1979): 1829–35.

24. H. Perez, J.A. Clegg, and S.R. Smithers, "Acquired Immunity to *Schistosoma mansoni* in the Rat: Measurement of Immunity by the Lung Recovery Technique," *Parasitology* 69 (1974): 349–59.

25. E.H. Sadun, F. von Lichtenberg, and J.I. Bruce, "Susceptibility and Comparative Pathology of Ten Species of Primates Exposed to Infection with *Schistosoma mansoni*," *American Journal of Tropical Medicine and Hygiene* 15 (1966): 705–18.

26. C. Peck, M. Carpenter, and A.A.F. Mahmoud, "The Role of Host Monocytes and Macrophages in Species-Dependent Resistance to *Schistosomiasis mansoni*," submitted for publication.

27. A. Sher, S.R. Smithers, and D. Mackenzie, "Passive Transfer of Acquired Resistance to *Schistosoma mansoni* in Laboratory Mice," *Parasitology* 70 (1975): 347–57.

28. A.I. Kassis, K.S. Warren, and A.A.F. Mahmoud, "Antibody-Dependent Complement-Mediated Killing of Schistosomula in Intraperitoneal Diffusion Chambers in Mice," *Journal of Immunology* 123 (1979): 1659–62.

29. A.A.F. Mahmoud, K.S. Warren, and P.A. Peters, "A Role for the Eosinophil in Acquired Resistance to *Schistosoma mansoni* Infection as Determined by Antieosinophil Serum," *Journal of Experimental Medicine* 142 (1975): 805–13.

30. A.A.F. Mahmoud, "Antieosinophil Serum," *American Journal of Tropical Medicine and Hygiene* 26 (suppl.) (1977): 104–12.

31. D.A. Dean, R. Wistar, and K.D. Murrell, "Combined in Vitro Effects of Rat Antibody and Neutrophilic Leukocytes on Schistosomula of *Schistosoma mansoni*," *American Journal of Tropical Medicine and Hygiene* 23 (1974): 420–28.

32. A.E. Butterworth, R.F. Sturrock, V. Houba, et al., "Antibody-Dependent Cell-Mediated Damage to Schistosomula in Vitro," *Nature* 252 (1974): 503–05.

33. A.A.F. Mahmoud, K.S. Warren, and G.T. Strickland, "Acquired Resistance to Infection with *Schistosoma mansoni* Induced by *Toxoplasma gondii*," *Nature* 263 (1976): 56–57.

34. A.E. Butterworth, M.A. Vadas, and J.R. David, "Mechanisms of Eosinophil Medicine Helminthotoxicity," in *The Eosinophil in Health and Disease*, ed. Adel A.F. Mahmoud and K. Frank Austen (New York: Grune and Stratton, 1980): 253–74.

35. J.W. Kazura, "Protective Role of Eosinophils," in *The Eosinophil in Health and Disease*, ed. Adel A.F. Mahmoud and K. Frank Austen (New York: Grune and Stratton, 1980): 231–52.

36. J.W. Kazura, M.M. Fanning, J.T. Blumer, et al., "Role of Cell-Generated H_2O_2 in Granulocyte-Mediated Killing of Schistosomula of *Schistosoma mansoni*," *Journal of Clinical Investigation* 67 (1981): 93–102.

37. E.C. Jong, A.A.F. Mahmoud, and S.J. Klebanoff, "Peroxidase-Mediated Toxicity to Schistosomula of *Schistosoma mansoni*," *Journal of Immunology* 126 (1981): 468–71.

38. A.E. Butterworth, D.L. Wassom, G.J. Gleich, et al., "Damage to Schistosomula of *Schistosoma mansoni* Induced Directly by Eosinophil Major Basic Protein," *Journal of Immunology* 122 (1979): 221–29.

39. P. Venge, R. Dahl., R. Hallgre, et al., "Cationic Proteins of Human Eosinophils and Their Role in the Inflammatory Reaction," in *The Eosinophil in Health and Disease*, ed. Adel A.F. Mahmoud and K. Frank Austen (New York: Grune and Stratton, 1980): 131–44.

40. G.R. Olds., J.J. Ellner, L.A. Kearse, Jr., et al., "Role of Arginase in Killing of Schistosomula of *Schistosoma mansoni*," *Journal of Experimental Medicine* 151 (1980): 1557–62.

41. R.H. Civil, K.S. Warren, and A.A.F. Mahmoud, "Conditions for Bacille Calmette-Guerin Induced Resistance to Infection with *Schistosoma mansoni* in Mice," *Journal of Infectious Diseases* 137 (1978): 550–55.

42. A.A.F. Mahmoud, P.A.S. Peters, R.H. Civil, et al., "In Vitro Killing of Schistosomula of *Schistosoma mansoni* by BCG- and *C. parvum*-Activated Macrophages," *Journal of Immunology* 122 (1979): 1655–57.

43. A.A. Majid, T.F. de C. Marshall, M.F. Hussein, et al., "Observations on Cattle Schistosomiasis in the Sudan, a Study in Comparative Medicine. I. Epizootiologic Observations on *Schistosoma bovis* in the While Nile Province," *American Journal of Tropical Medicine and Hygiene* 29 (1980): 435–41.

44. H.O. Bushara, A.A. Majid, A.M. Saad, et al., "Observations on Cattle Schistosomiasis in the Sudan, a Study in Comparative Medicine. II. Experimental Demonstration of Naturally Acquired Resistance to *Schistosoma bovis*," *American Journal of Tropical Medicine and Hygiene* 29 (1980): 442–51.

45. A.A. Majid, H.O. Bushara, A.M. Saad, et al., "Observations on Cattle Schistosomiasis in the Sudan, a Study in Comparative Medicine. III. Field Testing of an Irradiated *Schistosoma bovis* Vaccine," *American Journal of Tropical Medicine and Hygiene* 29 (1980): 452–55.

DISCUSSANT:

Nadia Nogueira

For nearly seventy years parasitologists have been trying to develop vaccination procedures against *Trypanosoma cruzi,* the agent of Chagas' disease.[1] Short of sublethal infection with live parasites, none of the experimental protocols has worked. Most of them have been aimed at raising antibodies in the expectation that these would afford protection.

On the basis of these findings, and of experimental evidence suggesting that cell-mediated mechanisms are involved in protective immunity,[2] we decided to study the basic phenomena in the interaction of this parasite with the host's immune system. We wanted to know the role of cell-mediated versus humoral immunity in protection, as well as the basic mechanisms by which the parasites interact with cells.

Figure 1 shows a scanning electron micrograph of a trypomastigote of *T. cruzi* being taken up by a normal mouse macrophage. Macrophages are important effector cells in immunity, and we started by investigating their role in this system. Trypomastigotes of *T. cruzi* phagocytized by normal macrophages are initially found in a phagocytic vacuole, but they soon lyse the vacuole and escape into the cytoplasm where they replicate. Normal macrophages therefore provide a favorable environment for the replication of the parasite. In contrast, when mice are sublethally infected with *T. cruzi,* and macrophages are obtained from their peritoneal cavity following challenge with heat-killed trypanosomes, the macrophages obtained resemble those shown in Figure 2. These are activated macrophages, quite different morphologically and functionally

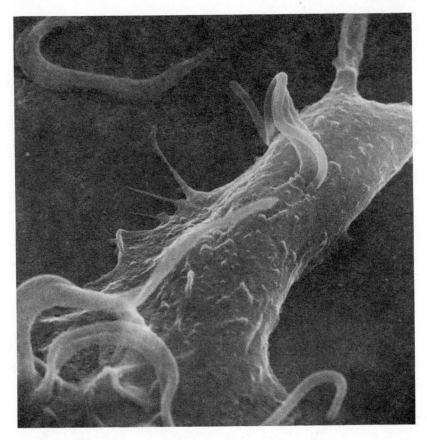

Figure 1. Scanning electron micrograph of a Y strain trypomastigote in early stage of phagocytosis by a normal mouse peritoneal macrophage. Magnification, × 4,125. Micrograph kindly taken by Dr. G. Kaplan.

from those in Figure 1. The most important difference for us is that these macrophages are now capable of efficiently killing the phagocytized parasites.[3] The mice from which the activated macrophages were taken possess strong protective immunity to challenge with the virulent blood forms of the organism (Figure 3). All the animals survive challenge with 10^5 blood forms of *T. cruzi* and display very low parasitemia levels. Age- and sex-matched control littermates display high levels of parasites in the blood and they die within twenty-one days. Figure 4 indicates that this protection can be passively transferred by T

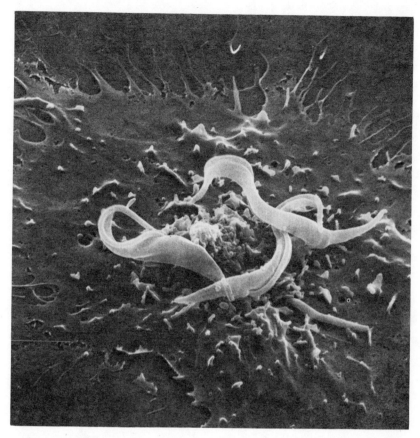

Figure 2. Scanning electron micrograph of Y strain trypomastigotes attached to the surface of a macrophage activated by in vitro infection with *Trypanosoma cruzi*. Notice the circumspherically spread appearance of the cell. Magnification, × 4,125. Micrograph kindly taken by Dr. G. Kaplan.

cell-enriched spleen cell populations from mice sublethally infected with *T. cruzi* two to four weeks earlier.[4]

For a detailed study of the effector cells mediating antitrypanosome activity we turned to an in vitro system, exposing the immune spleen cells in vitro to heat-killed trypanosomes and collecting the supernatants of such cultures.[5] Macrophages were then exposed to the mediators so generated and their trypanocidal activity was evaluated; the same state of activation may be achieved by exposing macrophages to such spleen cell factors (Figure 5). We determined that immune T cells are

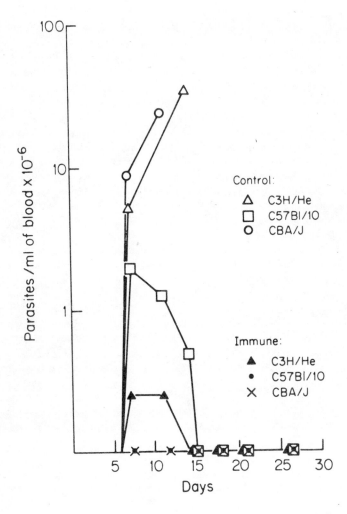

Figure 3. Course of parasitemia in control and immune mice challenged intraperitoneally with 10^5 trypomastigotes of the Y strain of *Trypanosoma cruzi*. Immune mice were obtained by sublethal infection four weeks earlier with 5×10^6 culture forms of the CL strain of *T. cruzi*.

SOURCE: Reproduced, by permission, from N. Nogueira, J. Ellis, S. Chaplan, et al., "*Trypanosoma cruzi*: *in Vivo* and *in Vitro* Correlation Between T-Cell Activation and Susceptibility in Inbred Strains of Mice." *Experimental Parasitology* 51 (1981), in press.

necessary to obtain the active supernatant fluids and that antigenic specificity is required at this stage. The lymphocyte products in turn are able to activate macrophages to a

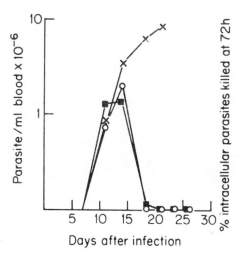

Figure 4. Course of parasitemia in C3H/HeJ mice infected with 10[4] Y strain trypomastigotes alone (x), plus 10[8] immune spleen cells (■) or 4×10[7] T cell-enriched (immunoglobulin-bearing depleted) immune spleen cells (o).

Source: Reproduced, by permission, from N. Nogueira, J. Ellis, S. Chaplan, et al., "*Trypanosoma cruzi: in Vivo* and *in Vitro* Correlation Between T-Cell Activation and Susceptibility in Inbred Strains of Mice," *Experimental Parasitology* 51 (1981), in press.

nonspecific microbicidal state. This data therefore provide evidence that, during the acute phase of the infection, cell-mediated immune mechanisms seem to be an important component in protection.

We have also been trying to unravel the biochemical nature of the intracellular killing mechanism of the activated macrophages. In this respect we have found that macrophage trypanocidal activity correlates closely with the ability of these cells to generate large amounts of H_2O_2 in a variety of experimental conditions.[6] This oxygen metabolite is highly microbicidal, as Seymour J. Klebanoff points out in his chapter in this volume. Trypanosomes are particularly susceptible because they lack catalase. We therefore think that oxygen intermediates, generated by activated macrophages, may be an important component of the microbicidal capacity of these cells.

When we turned to the investigation of the interaction of the blood forms of the parasite with macrophages we obtained quite unexpected results. These blood forms are much less phagocytized by macrophages than are the metacyclic

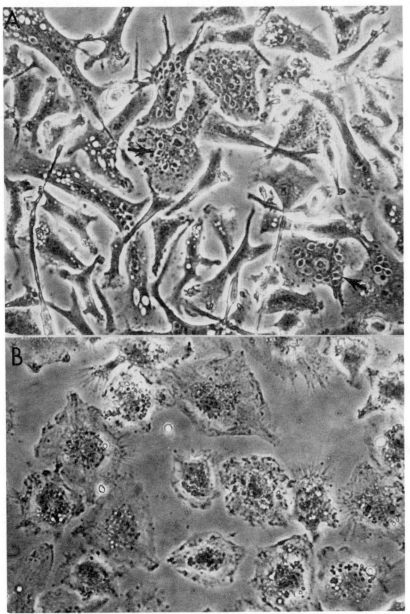

Figure 5. Phase-contrast micrographs of PP-induced macrophages seventy-two hours after infection with *Trypanosoma cruzi.* A = cells cultivated in medium alone for the entire time. B = cells cultivated in medium containing 25 percent of BCG-induced lymphokine. Magnification, × 600.

Source: Reproduced, by permission, from N. Nogueira and Z. Cohn, "*Trypanosoma cruzi: in Vitro* Introduction of Macrophage Microbicidal Activity," *Journal of Experimental Medicine* 148 (1978): 288–300.

trypomastigotes from cultures. A closer look revealed that this antiphagocytic effect could be overcome by trypsinizing the parasite surface. Parasites are then promptly ingested, survive quantitatively, and replicate within the cytoplasm of the normal macrophage, as do metacyclic trypomastigotes.

Another way this antiphagocytic effect may be obviated is by coating the parasite with specific antibodies of the IgG class. This involves phagocytosis mediated by the macrophage Fc receptor, which does not affect the parasite's ability to survive and multiply within normal macrophages.

The same two procedures enhance phagocytosis of blood forms by activated macrophages, and their intracellular fate is the same as that of the metacyclic trypomastigotes: intracellular killing and digestion. Table 1 shows that the IgG coating of parasites enhances their uptake manifold by both normal and lymphokine-activated macrophages. Within normal macrophages parasites survive and replicate; within activated macrophages they are killed more efficiently and at a faster rate than nonopsonized parasites.[7]

These experiments therefore suggest that blood-form trypomastigotes develop an evasion mechanism by which they avoid the macrophage activation system. They seem to possess on their surface a certain component that renders them non-

TABLE 1. FATE OF TRYPSINIZED AND OPSINIZED BLOOD FORM TRYPOMASTIGOTES OF *Trypanosoma cruzi* IN MACROPHAGE POPULATIONS (MACs)

Treatment of Macrophages	Treatment of Parasites		Parasites per 100 MACs at 72 Hours	Percent Infected MACs at 72 Hours
	Trypsin	Serum		
Medium alone	+	−	231	26
+ Lymphokine	+	−	40	11
+ Control supernatant	+	−	269	25
Medium alone	+	+	558	54
+ Lymphokine	+	+	0	0
+ Control supernatant	+	+	634	62

SOURCE: Reproduced, by permission, from N. Nogueira and Z. Cohn, "*Trypanosoma cruzi: in Vitro* Induction of Macrophage Microbicidal Activity," *Journal of Experimental Medicine* 148 (1978): 288–300.

phagocytizable. In the absence of macrophage activation the presence of specific antibodies confers no protection; it serves only as a vehicle to introduce more parasites into the macrophages where they replicate. By contrast, in the presence of macrophage activation, specific antibodies direct the parasite into the mononuclear phagocytes and potentiate their intracellular killing. These facts suggest that in vivo humoral immunity may also be efficient only in the presence of concomitant cell-mediated immunity.

More recently we have begun to investigate in detail the nature of this antiphagocytic factor and to identify immunologically relevant parasite surface components.[8] The approach we have taken is to identify parasite surface peptides by surface and internal labeling coupled with immunoprecipitation techniques.

Figure 6 shows the surface iodination profile of the blood form of the parasite before and after trypsinization. The major surface component of these forms, a 90,000 M_r glycoprotein, is completely removed by trypsinization of the parasite surface. This component is a homogeneous peptide, of isoelectric point 5, as seen in Figure 7 in two-dimensional gel electrophoresis; it may be associated with the antiphagocytic effect described earlier.

Figure 8 illustrates that this glycoprotein is absent from the surface of metacyclic trypomastigotes, which, in contrast, possess a surface peptide of 75,000 M_r of isolectric point 7.4. The peptide is not removed by trypsin treatment, nor do these forms show an antiphagocytic effect.

The immunoprecipitation pattern of the blood-form surface components obtained with mouse hyperimmune sera is quite similar to that obtained with a panel of IgG antibodies from patients with chronic Chagas' disease (Figure 10). These same antibodies fail to recognize any major intracellular peptide (labeled with ^{35}S-methionine). It is therefore likely that surface components may indeed be the relevant parasite antigens involved in the induction of protective immunity.

Thus immunology can provide parasitologists with the basic framework within which to plan a rational and effective approach toward the development of vaccination procedures.

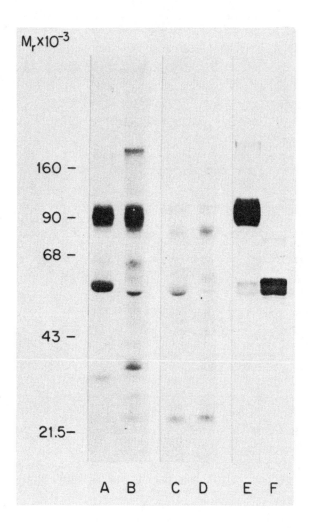

Figure 6. Effect of trypsin treatment of blood form trypomastigotes on the immunoprecipitated proteins. A = Y strain immunoprecipitation of ^{125}I-labeled, whole-cell lysate with hyperimmune mouse serum. B = Y strain, profile of ^{125}I-labeled, whole-cell lysate with human immune IgG. C = as in A, parasites trypsinized before iodination. D = as in B, parasites trypsinized before iodination. E = CL strain, [35S] methionine-labeled surface components immunoprecipitated with hyperimmune serum. F = as in E, parasites trypsinized before incubation with hyperimmune mouse serum.

SOURCE: Reproduced, by permission, from N. Nogueira, S. Chaplan, J. Tydings, et al., "*Trypanosoma cruzi*: The Surface Antigens of Blood and Culture Forms," *Journal of Experimental Medicine* 153 (1981): 629–39.

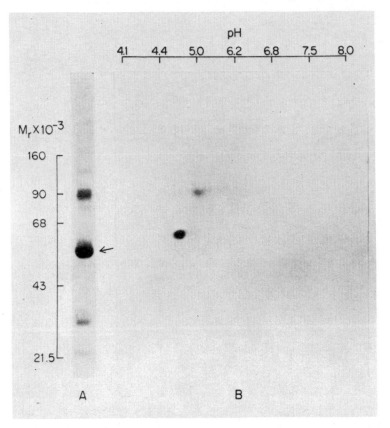

Figure 7. Two-dimensional gel electrophoresis of surface peptides of blood form trypomastigotes. A = one-dimensional profile of [³⁵S] methionine-labeled blood forms, CL strain, surface components immunoprecipitated with mouse hyperimmune serum. B = same samples run in two dimensions.

SOURCE: Reproduced, by permission, from N. Nogueira, S. Chaplan, J. Tydings, et al., "*Trypanosoma cruzi*: The Surface Antigens of Blood and Culture Forms," *Journal of Experimental Medicine* 153 (1981): 629–39.

This includes identification of the relevant antigens and immunization protocols directed toward enhancing one or another arm of the immune system.

On the other hand, parasitology can provide immunologists with fascinating models to study the immune system, as is the case with *T. cruzi,* which provides a valuable system with which to study cell-mediated immunity.

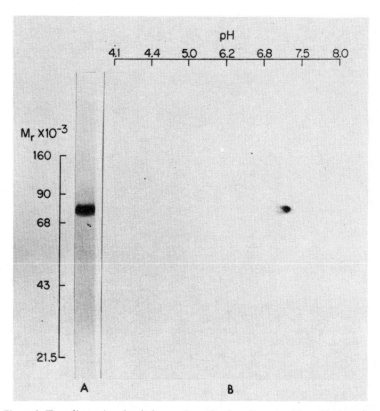

Figure 8. Two-dimensional gel electrophoresis of surface peptides of culture form trypomastigotes. A = one-dimensional profile of ^{125}I-labeled, whole-cell lysate of metacyclic trypomastigotes, CL strain. B = two-dimensional profile of the same sample in A.

SOURCE: Reproduced, by permission, from N. Nogueira, S. Chaplan, J. Tydings, et al., "*Trypanosoma cruzi:* the Surface Antigens of Blood and Culture Forms," *Journal of Experimental Medicine* 153 (1981): 629–39.

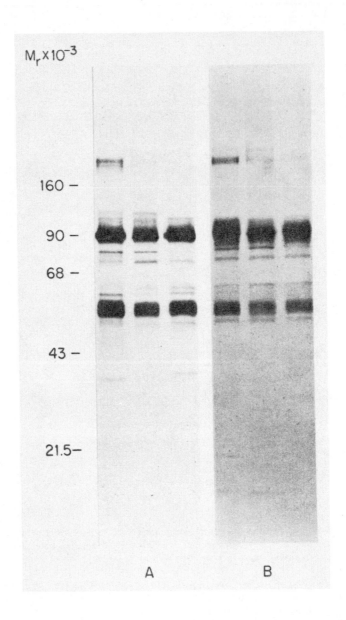

Figure 9. Immunoprecipitation profiles of [³⁵S] methionine-labeled blood forms with human immune IgG fractions. A = Y strain, human immune IgG fractions. B = CL strain, human immune IgG fractions.

SOURCE: Reproduced, by permission, from N. Nogueira, S. Chaplan, J. Tydings, et al., "*Trypanosoma cruzi*: The Surface Antigens of Blood and Culture Forms," *Journal of Experimental Medicine* 153 (1981): 629–39.

Notes

1. F.C. Goble, in *Immunity to Parasitic Animals*, ed. G. T. Jackson, R. Herman, and I. Singer (New York: Appleton-Century-Crofts, 1970): 597–689.

2. T. Pizzi, M. Rubio, and F. Knierim, "Immunologie de la Enfermedad de Chagas," *Boletin Chilean Parasitologia* 9 (1954): 35.

3. N. Nogueira and Z. Cohn, "*Trypanosoma cruzi*: Uptake and Intracellular Fate in Normal and Activated Cells," *American Journal of Tropical Medicine and Hygiene* 26 (1977): 194.

4. N. Nogueira, J. Ellis, S. Chaplan, et al., *Trypanosoma cruzi: in Vivo* and *in Vitro* Correlation Between T-Cell Activation and Susceptibility in Inbred Strains of Mice," *Experimental Parasitology* 51 (1981), in press.

5. N. Nogueira and Z. Cohn, "*Trypanosoma cruzi: in Vitro* Induction of Macrophage Microbicidal Activity," *Journal of Experimental Medicine* 148 (1978): 288–300.

6. C. Nathan, N. Nogueira, C. Juangbhanich, et al., "Activation of Macrophages *in Vivo* and *in Vitro* Correlation Between Hydrogen Peroxide Release and Killing of *Trypanosoma cruzi*," *Journal of Experimental Medicine* 149 (1979): 1056.

7. N. Nogueira, S. Chaplan, and Z. Cohn, "*Trypanosoma cruzi*: Factors Modifying Ingestion and Fate of Blood Form Trypomastigotes," *Journal of Experimental Medicine* 152 (1980): 447.

8. N. Nogueira, S. Chaplan, J. Tydings, et al., "*Trypanosoma cruzi*: the Surface Antigens of Blood and Culture Forms, *Journal of Experimental Medicine* 153 (1981): 629–39.

POPULATION BIOLOGY OF PARASITIC INFECTIONS*

Robert M. May

INTRODUCTION

Many contributions in this volume emphasize recent advances in immunology, molecular biology, and biochemistry, and the light they shed on the interactions between specific parasites and their individual hosts. This chapter, to the contrary, deals with the overall population dynamics of parasitic infections (with "parasite" broadly defined to include viruses, bacteria, fungi, protozoans, and helminths). That is, while other chapters point to new avenues opened up by our growing understanding of immune processes in individuals, this one focuses on complementary questions about "herd immunity."

Such discussion of the population biology of parasitic infections—their transmission, maintenance, and possible regulatory effects on their host populations—is important for several reasons.

First, it provides a general framework within which the vast array of information about parasitic infections may be organized in an orderly way, giving emphasis to the similarities and differences among the various parasites and identifying

* This work was supported in part by a National Science Foundation grant DEB77–01565.

the ecologically based patterns of relationship among epidemiological parameters (transmission rates, pathogenicity, lifespan of the parasite within a host, etc.). This framework, based on the overall ecology and epidemiology of the host-parasite association, is in some respects interestingly different from the conventional approach, which classifies parasites under the taxonomic headings of virus, bacterium, fungus, protozoan, helminth, and arthropod.[1-5]

Second, well-designed public health programs in general, and vaccination programs in particular, require an understanding of the process of infection both in individuals (as discussed elsewhere in this book) and in the population as a whole (as discussed here). Such understanding at the population level helps to explain why vaccination programs have effectively eradicated smallpox throughout the world, and poliomyelitis and diphtheria in developed countries, while in the latter they have been less successful against measles, whooping cough, malaria, and schistosomiasis.[6] This understanding makes it clear that there can be no single strategy or panacea to eradicate or control parasitic infections, but rather that different classes of infections require different kinds of approaches; this lesson has been learned the hard way in the related research area of pest control.[7]

Third, it has been remarked elsewhere in this book that the infusion of ideas from molecular biology and biochemistry into parasitology carries with it an infusion of talented young pre-doctoral and postdoctoral research workers who are helping to revitalize the discipline. Likewise, the deliberate injection of more evolutionary ecology and population biology into parasitological research will help to bring able researchers from those currently lively disciplines in its wake.

In what follows, I begin by outlining some of the main themes that have been developed in connection with the population biology of parasitic infections. These include distinctions between microparasites and macroparasites, and between infection and disease. The basic reproductive rate of a parasitic infection, threshold host densities, and the circumstances under which an infection may actually regulate the density of its host population are also discussed. To help make the presentation more concrete (and to help convince the skeptic), some applications of these ideas are then sketched. The examples include: regulation of populations of laboratory mice by specific viral or bacterial infections; interactions between forest insect populations and their microspordian protozoan or baculovirus infections; and rabies in fox populations in Europe. The chapter concludes with a discussion of the epidemiology of some parasitic infections in human populations, including analysis of vaccination programs against directly transmitted infections (particularly measles and whooping cough in Britain) and of control programs against indirectly transmitted infections (malaria and schistosomiasis).

SOME GENERAL THEMES

An increasing amount of attention is being given to the ways in which parasitic infections (broadly defined, as above) may regulate natural populations of animals. Such regulation may derive from the parasitic infection acting alone, or in conjunction with other regulatory mechanisms, such as predators or limited food supplies. R. M. Anderson and I have reviewed laboratory studies which show that various host populations are regulated by infectious diseases;[8-10] the corresponding evidence for natural populations is more scattered and anecdotal, but it seems likely that—at very least—many arthropod populations are controlled by parasitic infections. The interplay between disease and stresses induced by inadequate nutrition seems likely to be important in many instances; it certainly is for human populations, where the majority of infant deaths in underdeveloped countries appear to be due to childhood infections that are not lethal in the better nourished populations of developed countries.[11] Figure 1 bears witness to this large-scale interaction between disease and demography in human populations.

Recent work on the population biology of parasitic infections combines elements of ecological predator-prey theory (dealing with field, laboratory, and theoretical studies of the way prey populations may be regulated by their predators) with elements of classical epidemiology (which treats the transmis-

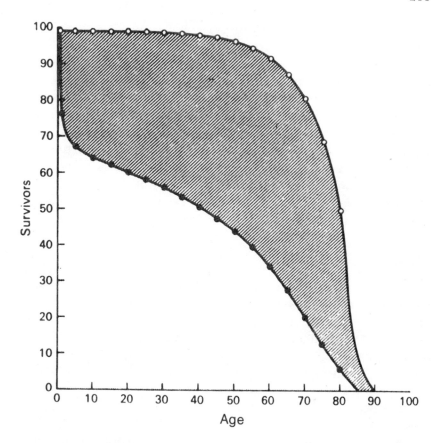

Figure 1. The probability, expressed as a percentage, of surviving to a given age in a prosperous industrial country (open circles) and in a poor developing country (solid circles).

Source: After D. J. Bradley, "Human Pest and Disease Problems: Contrast Between Developing and Developed Countries," in *Origins of Pest, Parasite, Disease and Weed Problems*, ed. J. M. Cherrett and G. R. Sagar (Oxford: Blackwell Scientific Publications, 1977): 329–46.

sion and maintenance of infections within a host population that has a constant magnitude, determined by other factors). Some of the conclusions about the circumstances under which parasitic infections may control the magnitude of their host population, at a steady level or in stable cycles, are outlined in the examples that follow. Much of this work is in flat contradiction to the conventional wisdom, set forth in many parasitology texts, that successful parasites inflict little or no harm on their hosts; the arguments usually presented in support of this belief are superficial, and often naively group selectionist.[12] As sketched in the next section, and analyzed in detail elsewhere,[13-18] the clumped or "overdispersed" distribution of many parasites—particularly helminths—among their hosts makes it possible for most hosts to be little affected by their (relatively low) parasite burden, while at the same time the overall density of the host population is indeed regulated by parasite-induced mortality. More generally, the complicated interplay between pathogenicity and transmissibility of parasites leaves room for many coevolutionary paths to be followed, with many endpoints.

Microparasites and Macroparasites

By using the term "parasitic infection" to embrace organisms from viruses to helminths and arthropods, we are encompassing a great variety of life forms and of associated population parameters. Broadly, however, two classes may be distinguished.[19] Most viruses, bacteria, and protozoans fall under the heading of *microparasites*, characterized by small size, short generation times, extremely high rates of direct reproduction within the host, and a tendency to induce immunity to reinfection in those hosts that survive the initial onslaught. The duration of infection is typically short in relation to the expected lifespan of the host, and is therefore of a transient nature. (There are of course many exceptions, of which the slow viruses are a particularly remarkable instance.)

For microparasites, a good description of the dynamical behavior of the infection may be obtained by considering the host population to be divided into several discrete classes of individuals—susceptible, latent, infectious, immune—with well-defined rate processes governing the flow among classes. To a rough approximation, no distinction need be made between infection and disease, and there is no attempt to discriminate among various levels of infection; people either are, or are not, infected. Mathematical models of microparasitic infections are therefore typically based on "compartmental" models of the kind illustrated schematically in Figure 2.

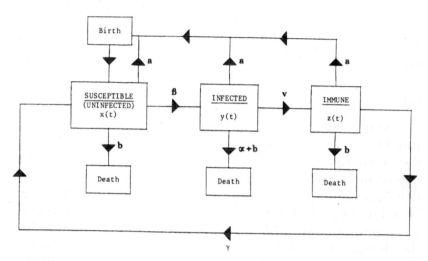

Figure 2. Diagrammatic flow chart for a directly transmitted infection described by a compartment model with susceptible (X), infected (Y), and immune (Z) hosts. The flow of individual hosts among compartments is controlled by a set of rate parameters: a, per capita birth rate; b, natural death rate of hosts; α, disease-induced mortality acting on infected hosts; v, recovery rate; β, transmission rate per encounter between susceptible and infected hosts; and γ, rate of loss of immunity.
SOURCE: After R. M. Anderson and R. M. May, "Population Biology of Infectious Diseases: Part I," *Nature* 280 (1979): 361–67.

Most parasitic helminths and arthropods (and some protozoans) come under the heading of *macroparasites.* Macroparasites tend to have much longer generation times than microparasites, and direct multiplication within the host is either absent or occurs at a low rate. The immune responses elicited by these metazoans generally depend on the number of parasites present in a given host and tend to be of relatively short duration. Macroparasitic infections therefore tend to be of a persistent nature, with hosts being continually reinfected.

For macroparasites the pathogenicity to the host, the rate of production of transmission stages of the parasite, and any resistance of the host to further infection all typically depend on the number of parasites present in a given host. A crude division of the host population into susceptible, infected, and immune classes is therefore not helpful, and a detailed description of the dynamics needs to deal with the full probability distribution of the parasites within the host population (that is, with the number of hosts harboring i parasites N(i), where i = 0, 1, 2, . . .). Figure 3, which is to be compared with Figure 2, depicts the essential structure of such models.

Infection Versus Disease

Notice that for microparasites infection and disease are roughly synonymous, whereas for macroparasites a significant distinction can generally be made between infection (harboring one or more parasites) and disease (harboring a parasite burden large enough to be pathogenic to the host). This distinction between infection and disease (or, equivalently, between prevalence and intensity of infection) has important implications for the design of control programs; one may choose to target chemotherapy against those hosts with the heaviest parasite burdens, which in many situations can dramatically reduce the incidence of disease while having little or no effect on the prevalence of infection. Moreover, as already mentioned, many parasitic infections are characterized by very overdispersed distribution among the host population, with most of the parasites

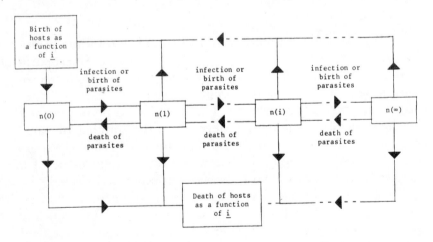

Figure 3. Diagrammatic flow chart for a directly transmitted infection based on a model with compartments for the number of hosts, N(i), harboring i parasites (i = 0, 1, 2, . . .). The model has a structure similar to, but obviously more complex than, that of Figure 2.
SOURCE: After R. M. May and R. M. Anderson, "Population Biology of Infectious Diseases: Part II," *Nature* 280 (1979): 455–61.

"clumped" in relatively few hosts; in these circumstances, programs targeted at these few, highly burdened hosts can be much cheaper than ambitious programs aimed at eradication, yet can be almost as effective in reducing human suffering. These ideas go back to W. G. Smillie's work on the control of hookworm infections in the southern United States,[20] and have recently been forcefully revived by K. S. Warren.[21,22]

In contrast to the wider definition employed in this chapter, most medical people understand "parasitology" to refer to protozoans and helminths. The preceding discussion suggests this is to lump together two broadly dissimilar classes of organisms—protozoans typically being microparasites, and helminths typically being macroparasites—having qualitatively different relations between infection and disease.

Needless to say, the distinction between microparasites and macroparasites is not hard-and-fast, but rather represents two ends of a continuum. Thus, for many protozoan and other "microparasitic" infections, detailed study of immune processes suggests them to be dependent on the intensity of infection and/or upon challenge by continual reinfection, so that some mathematical description intermediate between the schematizations of Figures 2 and 3 is needed. In particular, this seems to be the case for malaria.[23-25]

Basic Reproductive Rate of the Parasite

For a parasitic infection, the basic reproductive rate, R, may in general be defined as the expected number of secondary infections produced within the infectious lifetime of one newly introduced infected host. Clearly the infection is capable of establishing itself within the host population if, and only if, $R > 1$.

This notion of the basic reproductive rate is central to a discussion of the overall population biology of parasitic infections. The exact expression for R in terms of pertinent epidemiological parameters depends, however, on the dynamical details of the specific infection. For example, expressions for R will usually be more complicated for indirectly transmitted parasitic infections than for directly transmitted ones. More full discussions, complete with explicit formulae for R, are given by K. Dietz,[26] Anderson,[27] and Anderson and myself.[28,29]

Threshold Host Density

Many infections (particularly directly transmitted microparasitic infections) can persist within a host population only if it exceeds a certain critical, or "threshold," density, N_T. This phenomenon, first noted by W. O. Kermack and A. G. McKendrick,[30] suggests interesting correlations between the behavioral ecology of animal populations and the parasitic infections that can be maintained within them. For example, those directly transmitted microparasites that require high host densities in order to persist should be more commonly associated with animals that exhibit herd or shoaling behavior, or that breed in large colonies; empirical evidence in support of these ideas comes from the abundance of directly transmitted viral and bacterial infections in modern human societies, large herds of ungulates, breeding colonies of sea birds, and the social insects.[31] Conversely, those diseases with direct life cycles that do persist within low-density host populations should possess distinctive characteristics, such as long-lived infective stages, failure to induce lasting immunity, or the ability to persist within the host for very long times.

For many directly transmitted infections, R and N_T are related by the simple expression

$$R = N/N_T. \tag{1}$$

Here N is the host population magnitude or density. Equation (1) makes explicit the equivalence between the criteria $R > 1$ and $N > N_T$ for maintenance of the infection. For many indirectly transmitted infections, however, the relation corresponding to equation (1) is more complicated. Some other infections, such as many sexually transmitted diseases, fail to exhibit the threshold phenomenon.[32,33]

Persistence and Eradication of Parasitic Infections

Given the current interest in vaccines against malaria and other parasitic infections, it is worth noting the conditions for eradicating them. Basically, the requirement is that the number of susceptible hosts should be reduced to the point where R > 1. If a proportion, p, of the host population is vaccinated at or near birth (and revaccinated at appropriate intervals if the induced immunity is not lifelong), the effective value of R is usually reduced to $R' = R(1-p)$, and the correspondingly rough criterion for eradication of the infection is

$$p > 1 - 1/R. \tag{2}$$

This rough criterion, which is discussed more fully elsewhere,[34-37] shows that eradication requires almost 100 percent of the population to be protected if the basic reproductive rate of the parasite infection is high (R>>1).

The foregoing ideas will now be illustrated by an eclectic collection of examples.

GREENWOOD'S LABORATORY MICE

In a classic series of laboratory experiments, M. Greenwood and collaborators studied the dynamical behavior of populations of mice infected with the mouse pox virus ectromelia, and with the bacterium *Pasteurella muris*[38,39] (see also the work of F. Fenner[40,41]). The experimental design had many simplifying features, enabling attention to be focused on the host-parasite interaction: the cage space available to the mouse population was adjusted to keep the population density constant as absolute levels changed (thus eliminating density-dependent constraints on population growth); and uninfected adult mice were introduced at a specified rate (thus avoiding time lags and other complications attendant upon recruiting to the population by natural birth processes).

For such "microparasitic" viral and bacterial infections, the mouse population may be divided into the number that are susceptible, infected, and immune, denoted by X, Y, and Z, respectively. The mathematical model corresponding to the compartmental scheme indicated in Figure 2 is thus:

$$dX/dt = A - bX - \beta XY - \gamma Z, \tag{3}$$
$$dY/dt = \beta XY - (b+\alpha+v)Y, \tag{4}$$
$$dZ/dt = vY - (b+\gamma)Z. \tag{5}$$

Here the changes in X, Y, Z (that is, dX/dt, dY/dt, dZ/dt) are given as the differences between losses and gains (corresponding to the flows depicted in Figure 2). Specifically, A is the rate of introduction of new adult mice, b is the per capita death rate in the disease-free population, α is the per capita disease-induced mortality rate, v is the per capita recovery rate (into the immune class, Z), and γ is the rate at which immunity is lost ($\gamma = 0$ for lifelong immunity). The most debatable assumption is that infection is acquired at a rate proportional to the number of encounters between susceptible and infected mice, βXY (with β a transmission parameter).

Of this plethora of parameters, the introduction rate A is determined by the experimenter, while b, α, v, and γ can be determined from other studies independent of these population-level experiments. The transmission parameter β remains difficult to estimate as it depends on a variety of behavioral and epidemiological factors.

For a host-parasite system described by equations (3)–(5), the basic reproductive rate of the infection can be shown to be

$$R = \frac{\beta A}{b(\alpha+b+v)}. \qquad (6)$$

Thus the infection will be maintained if A is sufficiently large, $A > A_T$, with $A_T \equiv b(\alpha+b+v)/\beta$. If $R > 1$ $(A > A_T)$, the population will settle to a disease-regulated value of N^* that depends linearly on A:

$$N^* = a_1 A + a_2. \qquad (7)$$

Here a_1 and a_2 depend in a messy way on α, b, v, γ, and β.[42] If $R < 1$ $(A < A_T)$, the infection cannot be maintained and the mouse population will settle to the disease-free equilibrium, balanced between immigration and death: $N^* = A/b$.

For *Pasteurella muris*, Anderson and I estimate that b $\simeq$ 0.006, $\alpha \simeq 0.06$, v $\simeq 0.04$, and $\gamma \simeq 0.02$ (all in units of days^{-1}); β remains to be determined from the population experiments.[43] The solid dots in Figure 4a show the eventual equilibrium

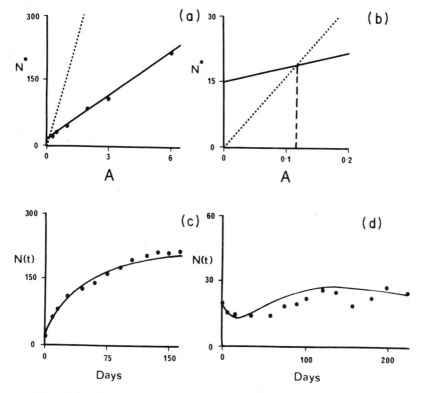

Figure 4. Population dynamics of *Pasteurella muris* in colonies of laboratory mice. (a): relationship between the equilibrium number of mice, N*, and the daily rate of input of susceptible mice, A. Solid dots are observed levels and the solid line is the best linear fit, equation (7); the dashed line shows the estimated relationship between N* and A in the absence of the disease (the slope is l/b). (b): enlargement of the portion of (a) where solid and dashed lines intersect, determining the threshold level of immigration, A_T, below which the disease will not persist. (c), (d): growth of mouse colonies harboring the disease, from an initial population of twenty mice, for A = 6.0 and 0.33, respectively. Again solid dots are the experimental data and solid lines the theoretical predictions described in the text.

SOURCE: After R. M. Anderson and R. M. May, "Population Biology of Infectious Diseases: Part I," *Nature* 280 (1979): 361–67.

population of mice, N*, obtained in the experiments of Greenwood and coworkers, as a function of the introduction rate A; they ran experiments for A = 6, 3, 2, 1, 0.5, 0.33 mice/day.[44] These data agree well with the linear relationship between N* and A predicted by equation (7). We estimate that the disease could not be maintained if the introduction rate were held below 0.11 mice/day (Figure 4b).

Moreover, the slope of the straight line fitted to the data points in Figure 4a gives an estimate of the one remaining parameter, β. With this value of β we can now use equations (3)–(5) to get a parameter-free prediction of the temporal development of the infection for any initial number of mice N(O) and introduction rate A. Two such fits between theory and data are shown in Figures 4c and 4d, for A = 6.0 and A = 0.33, respectively. Note the propensity to damped oscillations at relatively small A values. Bearing in mind the complete absence of adjustable parameters, both the fits are extremely encouraging.

A similar analysis may be carried out for the data obtained by Greenwood and colleagues for ectromelia infections.[45] Here $\alpha \simeq 0.042$, $v \simeq 0.014$, and $\gamma \simeq 0$, all in units of days^{-1}; the experiments were run only for a single introduction rate, A = 3 mice/day. Figure 5 (analogous to Figures 4c and 4d) thus shows a one-parameter fit (in terms of the parameter β, chosen to be $\beta \simeq 0.0013$ days^{-1}) between the data and the theory.

Several points emerge from the analysis of these experi-

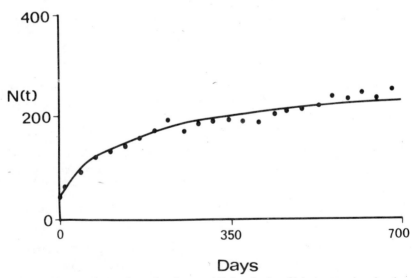

Figure 5. Population dynamics of ectromelia in colonies of laboratory mice, showing the growth of a mouse colony harboring the disease from an initial population of forty-five mice. Dots and solid curve as described in Figure 4c and d. Details are given in the text.

SOURCE: After R. M. Anderson and R. M. May, "Population Biology of Infectious Diseases: Part I," *Nature* 280 (1979): 361–67.

ments. First, they show that microparasitic infections can indeed regulate their host population. Second, Figure 4a illustrates the threshold phenomenon. Third, and possibly most significant for our present discussion, the agreement between theory and experiment provides confidence in the use of simple, deterministic models for studies of the overall population biology of parasitic infections.

The foregoing work is presented more fully in the paper by Anderson and myself.[46]

PARASITIC INFECTIONS OF FOREST INSECTS

Invertebrates do not appear to exhibit acquired immunity to the agents of infectious diseases; when invertebrates recover, rather than die, they generally pass back into the susceptible class. This feature makes the analysis of the dynamics of invertebrate host-parasite systems somewhat simpler than is the case for those vertebrate systems characterized by the presence of acquired immunity.

Recent studies draw together theoretical analysis with extensive compilations of empirical data, to suggest that many invertebrate populations may be regulated by viral, bacterial, or protozoan infections.

In particular, microspordian protozoan and baculovirus infections of many forest insect species appear to fulfill the conditions necessary for them to be influential in the regulation of their insect host populations. The free-living transmission stages of many of these parasites are relatively long-lived, which complicates the analysis; broadly, it can be shown that such infections with long-lived transmission stages can regulate the host population to a steady level, or in stable cycles. Stable cycles tend to ensue when the infection is highly pathogenic, in a host population whose intrinsic growth rate is relatively low. Thus microspordian and baculovirus infections that are very pathogenic and have long-lived transmission stages tend to induce population cycles in univoltine forest insect species; these cycles typically will have periods of around five to twelve years.

As a specific example, Figure 6A shows observed changes in the abundance (solid line) of the larch bud moth, *Zeiraphera diniana*, in the European Alps, and in the prevalence (dashed line) of infection with a granulosis virus. Figure 6B shows the corresponding theoretical results, with all the parameters (except the transmission parameter, β) estimated from other data, independent of the population data in Figure 6A. Again, the agreement is encouraging.

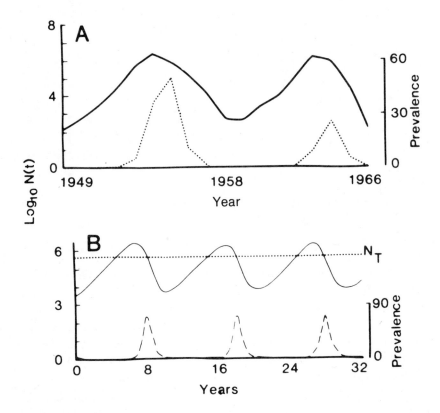

Figure 6. A: observed changes in the abundance, N(t) (plotted logarithmically, solid line) of the larch bud moth, *Zeiraphera diniana*, in the European Alps, and in the prevalence (expressed as a percentage and plotted linearly, dashed line) of infection with a granulosis virus. B: corresponding theoretical results for the host abundance and the prevalence of infection, as discussed briefly in the text.

SOURCE: After R. M. Anderson and R. M. May, "Infectious Diseases and Population Cycles of Forest Insects," *Science* 210 (1980): 658–61, which should be consulted for details.

Such discussions of the possible regulation of invertebrate populations by parasitic infections are interesting both for the light they may shed on observed phenomena in natural populations and for their possible application to calculate the rate at which a virus or other pathogen must be artificially introduced if it is to be effective in the control (or extinction) of a population of insect pests.

The work summarized in Figure 6 is presented more fully elsewhere,[47] while general aspects of the population biology of parasitic infection of invertebrate hosts is pursued in a monograph by Anderson and myself.[48]

FOX RABIES IN EUROPE

Another concrete application of the above ideas is to the interaction between rabies virus and fox populations in Europe. In those regions where rabies is now endemic in fox populations, a model of the general kind illustrated by Figure 2 may be used to analyze the dynamics of the interaction.

In detail, the model is more complicated than that depicted in Figure 2: a latent class must be added (for fox rabies the latent period of around thirty days is longer than the infectious period of around three to ten days); and the birth and natural death rates in fox populations tend to be density dependent. As before, the various ecological and epidemiological parameters can be estimated independent of the overall population data; the transmission parameter β—always a difficulty—can be obtained from estimates of the threshold fox density needed to maintain rabies infections, which is estimated to be about 1 fox/km². The resulting model predicts the fox population to be regulated by the infection, exhibiting stable cycles with periods in the range of three to five years, and with an average rabies prevalence of around 4 percent. These results accord with the observed population data: a striking feature of rabies epidemiology in Europe and North America is the regular three-, four-, or five-year cycles in fox density in most regions (including those where "wildlife's four-year cycle" in mice, vole, and lemming populations is not found), and prevalence levels in the range of 3–7 percent.

This approach subsumes all those complexities of fox behavior and ecology that influence the overall contact rate between rabid and susceptible foxes into a simple "βXY" term. It is pleasing both that this simple, aggregated model is in qualitative agreement with much more complicated and detailed analyses,[49] and that it explicitly and successfully relates the threshold fox population density (1 fox/km^2) to the average contact rate between foxes (family groups meet, on average, every four to six days).

Such theoretical models may be modified to evaluate the relative efficacy of different control strategies: culling or vaccination (oral vaccines delivered in baits), or combinations of these two. A full exposition of this work, and the background to it, is given by Anderson and coworkers.[50]

PARASITIC INFECTIONS IN HUMANS

Childhood Infections and Vaccination Programs

For many directly transmitted microparasitic infections, the basic reproductive rate of the parasite can be estimated from the approximate relation[51,52]

$$R = 1 + L/A. \tag{8}$$

Here L is the expected lifespan in the host population (around seventy years in human populations in developed countries, and significantly less in many underdeveloped countries), and A is the average age at which individuals acquire the infection. For a more exact analysis it is necessary to acknowledge that death rates and rates at which infections are acquired tend to be age-dependent; the resulting expression for R is

$$R = \frac{\int_0^\infty \exp\left[-\int_0^a b(v)dv\right]da}{\int_0^\infty \exp\left[-\int_0^a \{b(v) + \lambda(v)\}dv\right]da}. \tag{9}$$

Here b (a) is the per capita mortality rate, and λ (a) the per capita rate at which susceptible individuals acquire infection, at age a.

Table 1 sets out values of R, estimated by using equation (8), and published data pertaining to the age at first infection, A, for a variety of childhood infections of humans.

The fraction of a population that must be vaccinated (essentially at birth) in order to eradicate an infection, whose basic reproductive rate (in the prevaccinated population) is R, is given by equation (2). It is clear that the larger R, the larger the proportion that must be protected. Referring to Table 1, we see that diphtheria and poliomyelitis, the infections for which such vaccination programs have been successful, are characterized by relatively small R values, while measles and whooping cough, for which extensive vaccination programs have not succeeded in effective eradication, have relatively large R values. Although the design and evaluation of vaccination programs is

TABLE 1. ESTIMATES OF THE BASIC REPRODUCTIVE RATE (R)
FOR VARIOUS CHILDHOOD DISEASES

Disease	R
Measles	
Great Britain	12–18
North America, various locations	6–16
Ghana; Nigeria	14–16
Jordan	19
Whooping Cough	
Great Britain; United States	10–12
Chicken Pox	
North America, various locations	7–10
Mumps	
North America, various locations	4–7
Rubella	
Great Britain; Federal Republic of Germany	6
Poliomyelitis	
United States; Netherlands	6
Diphtheria	
North America, various locations	3–6
Scarlet Fever	
North America, various locations	4–6

SOURCE: The figures are mostly averages condensed from the much more detailed and extensive compilation, with references, in R. M. Anderson and R. M. May, "The Epidemiology of Directly Transmitted Infectious Diseases: Control by Vaccination," paper submitted for publication.

clearly a vastly more complicated matter than indicated by the foregoing crude analysis, the broad insights afforded by Table 1 in conjunction with equation (2) would appear to be a valuable starting point. There is currently considerable interest in the development of vaccines against hepatitis A and B viruses and cytomegaloviruses; it will be helpful to have concomitant serological studies aimed at estimating the R values for these infections.

More generally, if the average age at which individuals are vaccinated is V (rather than vaccination being essentially at birth, V = 0), eradication of the infection requires that the proportion, p, embraced by the vaccination program exceed

$$p > \frac{1 + V/L}{1 + A/L}. \tag{10}$$

Here A represents the average age at which individuals acquire infection in the prevaccinated population; in the limit $V \to 0$, equation (10) reduces back to equations (2) and (8). Like equation (8), equation (10) is an approximation (exact only when all rate processes are age independent), and the exact criterion involves appropriate integrals (after the style of equation [9]) over age-specific schedules of vaccination, mortality, and acquisition of infection. Application of this analysis to the extensive data available for the incidence of, and vaccination against, measles in Great Britain shows that, for the prevaccination cohort of individuals born in 1956, R $\simeq$ 16.0, whereas the vaccination of 57 percent of the 1970 cohort (at an average age 2.2 years) lowered this only to R $\simeq$ 12.8. Anderson and I estimate that if the average age of vaccination were kept at 2.2 years, it would be necessary to vaccinate 96 percent of each cohort to eradicate measles in Britain, and that even the optimal policy of vaccinating at an average age close to, or less than, 1 year would require that 94 percent of each cohort be vaccinated.[53] A corresponding analysis for whooping cough shows the prevaccination 1940 cohort to have R $\simeq$ 16.3, and the 1970 cohort (81 percent vaccinated at an average age of 1.7 years) to have R $\simeq$ 6.3; this represents a substantially greater impact on R than for measles, but is still far short of the R < 1 required for eradication. Again it can be estimated that eradication of whooping cough in Britain would require vaccination of 96 percent of each cohort at the prevailing average of 1.7 years, or, alternatively, of slightly less than 95 percent if the average age at vaccination were reduced to just below age 1.

Figure 7 shows data pertaining to the incidence of, and vaccination against, measles and whooping cough in England and Wales from 1940 to 1979. In addition to the estimates outlined already, mathematical models of the kind depicted in Figure 2 can explain other dynamical features of these infections, such as the fairly regular periodicities in outbreaks of measles and whooping cough.[54] Analysis of the data illustrated in Figure 7, and further data stretching back to the turn of the century, shows the average interepidemic period to have been 2.2 years for measles, lengthening after the advent of vaccina-

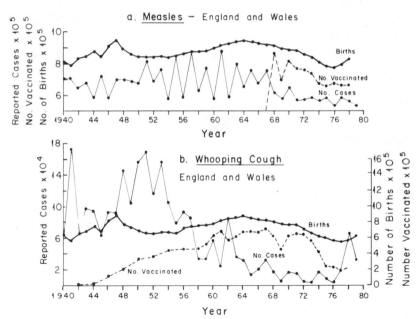

Figure 7. Reported cases of (a) measles and (b) whooping cough in England and Wales from 1940 to 1979. The figures show the total number of births (thick line), number of reported cases (thin line), and number of people vaccinated (dashed line) each year

Source: After R. M. Anderson and R. M. May, "The Epidemiology of Directly Transmitted Infectious Diseases: Control by Vaccination," submitted for publication.

tion to 2.6 years. The corresponding figures for whooping cough are 2.8 years before, and 3.5 years after, vaccination programs were begun. Both the absolute values of these interepidemic periods (2.2 years and 2.8 years), and the trends after vaccination was introduced (periods lengthening), are in quantitative agreement with theoretical estimates (which predict the interepidemic period in terms of the independently estimated quantities R, L, and the duration of the latent and infectious intervals).

The material in this section is a brief summary of work reviewed at length elsewhere.[55]

Schistosomiasis

As discussed earlier, and as the comparison between Figures 2 and 3 indicates, the overall population dynamics of macroparasitic infections can be more complicated than those of microparasites, and the range of possible public health measures can become correspondingly more diverse. This is particularly true when the macroparasite is transmitted indirectly.

Schistosomiasis is an indirectly transmitted macroparasitic infection of man that has been the subject of much field, laboratory, and theoretical work,[56-60] Many studies have been concerned with evaluating the relative effectiveness of different ways in which the overall transmission cycle can be interrupted: latrines can reduce the number of eggs, and thence miracidia, put into the aqueous environment; molluscicides may reduce the populations of intermediate host snails; safe water supplies may reduce contacts with cercariae to some extent; chemotherapy will reduce worm burdens in humans, and thence diminish egg output. Such studies aim to identify those biological features of the schistosome life cycle that are important to the overall transmission process, and then to combine this biology with economic considerations, to determine the control strategy that is "optimal" according to some criterion or other.

The fact that schistosomes—like most macroparasites—are distributed among their human host population in a highly clumped or overdispersed way is important, both for the basic population dynamics of the host-parasite interaction and for the evaluation of public health programs. Until recently the models for schistosomiasis assumed adult helminths to be distributed independently randomly among their human hosts; the pronounced degree of schistosome clumping, documented for human and other animal populations, significantly alters the transmission dynamics.[61] Moreover, as emphasized by Warren,[62] J. A. Walsh and Warren,[63] and others,[64] such clumping puts a premium on chemotherapy targeted against people with above-average worm burdens; this strategy is cost-effective, whether the ultimate aim is control of disease (by removing only those worm burdens high enough typically to be pronouncedly pathogenic), or the much more ambitious aim of eradicating infection.

Hookworm

Noting the overdispersed distribution of hookworms among human hosts, Smillie earlier suggested that public health programs be directed primarily at those hosts with relatively high hookworm burdens.[65]

More recently, Anderson has applied population models (of the general kind depicted in Figure 3) to this directly transmitted macroparasitic infection.[66] He has shown that data on age-specific prevalence and intensity of hookworm infection in human populations can be well explained by such relatively simple models; on this basis one can make quantitative estimates of the outcome of a specified control program.

Malaria

Modifying Ross's earlier model to incorporate the effects of the parasite's latent period in the mosquito, G. Macdonald showed that the basic reproductive rate for malarial parasites in man[67] is

$$R = \frac{ma^2 e^{-\mu T}}{r\mu}. \tag{11}$$

In this expression (which glosses over several complications), m is the number of mosquitoes per human host, a the biting rate (the average number of bites on a man per mosquito per day), μ the mosquito death rate, T the latent period in the mosquito, and r the recovery rate for infected humans. This rough result has been influential in suggesting that attacks directed at adult mosquitoes (that is, aimed at increasing μ) will, by virtue of the exponential dependence on μT, be likely to be more effective than attacks directed at larvae (that is, aimed at decreasing m). Complications, due essentially to the effects of spatial heterogeneity, have undercut practical applications of this simple formula.[68]

Equation (11) and its more elaborate descendants[69] are based on descriptions of the protozoan malarial species as indirectly transmitted microparasites. But the degree of acquired immunity to malarial infection possessed by many adults in endemic regions appears to depend, inter alia, on continued reinfection in order to "boost" immunity; mathematical models incorporating this important ingredient are, in a sense, in a category intermediate between "microparasites" and "macroparasites" (the models having structure intermediate between Figures 2 and 3).

The epidemiological consequences of these complexities in herd immunity are illustrated in Figure 8, which presents evidence showing that a temporary reduction in the malarial transmission rate ends up producing an increase in the prevalence of malaria among adults. More generally, Figure 9 shows curves for the prevalence of malaria as a function of age in four regions: curve D, corresponding to the highest transmission intensity, gives the highest prevalence among children, but the lowest among older people. Figure 10 shows how the main features of the data in Figure 9 can be exhibited by a mathematical model in which maintenance of acquired immunity requires relatively frequent reinfection.[70] The implications of this work for public health programs are distilled in Figure 11, which shows the net prevalence of malaria in adults (summed over all individuals older than age twenty), as a function of the transmission intensity, h (which depends on such things as m and a, as roughly indicated by Equation [11]). A reduction in h from very high values may actually increase the prevalence of malaria among the adult population, although the children will almost always benefit.

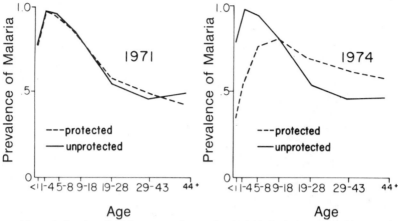

Figure 8. Graph on left shows prevalence of malarial infection versus age in years in an endemic area in 1971 before intervention; there is no difference yet between the two village groups (solid and broken lines). In 1972 and 1973 there was massive intervention in one of the groups, which was halted for adults in 1974. Graph on right shows prevalence of infection versus age in years in 1974. Adults in the once-protected group (broken line) show significantly higher prevalences than those in the comparative villages (solid line) in which there was no intervention.

SOURCE: After J. L. Aron and R. M. May, "The Population Dynamics of Malaria," in *Population Dynamics of Infectious Diseases*, ed. R. M. Anderson (London: Chapman and Hall, 1981). Data from R. Cornille-Brögger, H. M. Mathews, J. Storey, et al., "Changing Patterns in the Humoral Immune Response to Malaria Before, During, and After the Application of Control Measures: A Longitudinal Study in the West African Savanna," *Bulletin of the World Health Organization* 56 (1978): 579–600.

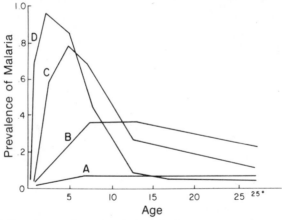

Figure 9. Prevalence of acute malaria infections versus age in years, in stable indigenous populations, for differing levels of endemicity. A: low endemicity; B: moderate endemicity; C: high endemicity; D: hyperendemicity.

SOURCE: After J. L. Aron and R. M. May, "The Population Dynamics of Malaria," in *Population Dynamics of Infectious Diseases*, ed. R. M. Anderson (London: Chapman and Hall, 1981); adapted from M. F. Boyd, "Epidemiology of Malaria: Factors Related to the Intermediate Host," in *Malariology*, ed. M. F. Boyd (Philadelphia: W. B. Saunders, 1949): 551–607.

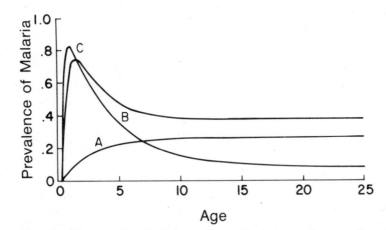

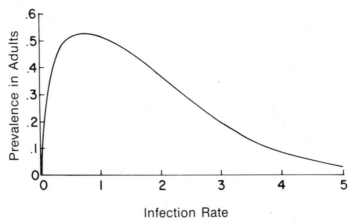

Figure 10. Theoretical results for prevalence of malaria as a function of age for three different transmission rates. The curves are derived from a "microparasitic" compartmental model in which, however, the rate of loss of immunity depends on the malaria transmission rate, h: (A) h = 0.1 yr⁻¹ (relatively low transmission rate); (B) h = 2 yr⁻¹; (C) h = 4 yr⁻¹ (relatively high transmission rate). The features of these curves, particularly the "crossing over" phenomenon, are as for the data summarized in Figure 9.
SOURCE: After J. L. Aron and R. M. May, "The Population Dynamics of Malaria," in *Population Dynamics of Infectious Diseases*, ed. R. M. Anderson (London: Chapman and Hall, 1981), in which the detailed derivation of this figure is presented.

Figure 11. The prevalence of malaria among adults over age twenty is shown as a function of infection rate, h (plotted in units of yr⁻¹), under the same assumptions about the rate of loss of immunity as in Figure 10. The implications of this curve for public health and control programs are briefly discussed in the text.
SOURCE: After J. L. Aron and R. M. May, "The Population Dynamics of Malaria," in *Population Dynamics of Infectious Diseases*, ed. R. M. Anderson (London: Chapman and Hall, 1981), in which a full derivation and discussion is given.

These and other aspects of models for the population biology of malarial infections are reviewed in detail by J. L. Aron and myself.[71]

CONCLUSION

My main messages were spelled out in the introduction to this chapter and in the section on General Themes. The subsequent examples sketched applications treated at length elsewhere.

NOTES

1. N. J. T. Bailey, *The Mathematical Theory of Infectious Diseases,* 2nd ed. (New York: Macmillan Co., 1975).

2. K. Dietz, "Transmission and Control of Arbovirus Diseases," in *Epidemiology,* ed. D. Ludwig and K. L. Cooke (Philadelphia: Society for Industrial and Applied Mathematics, 1975): 104–21.

3. ———, "The Incidence of Infectious Diseases Under the Influence of Seasonal Fluctuations," in *Mathematical Models in Medicine: Lecture Notes in Biomathematics,* vol. 11, ed. J. Berger, W. Buhlen, R. Repges, et al. (Berlin: Springer Verlag, 1976): 1–15.

4. R. M. Anderson and R. M. May, "Population Biology of Infectious Diseases: Part 1," *Nature* 280 (1979): 361–67.

5. R. M. May and R. M. Anderson, "Population Biology of Infectious Diseases: Part II," *Nature* 280 (1979): 455–61.

6. R. M. Anderson and R. M. May, "The Epidemiology of Directly Transmitted Infectious Diseases: Control by Vaccination," submitted for publication.

7. G. Conway, "Man versus Pests." in *Theoretical Ecology: Principles and Applications,* 2nd ed., ed. R. M. May (Oxford: Blackwell Scientific Publications; Amherst, Massachusetts: Sinauer, 1981): 356–86.

8. R. M. Anderson and R. M. May, "Regulation and Stability of Host-Parasite Population Interactions: I. Regulatory Processes," *Journal of Animal Ecology* 47 (1978): 219–47.

9. ———, "Population Biology. I" (See note 4).

10. ———, "The Population Dynamics of Microparasites and Their Invertebrate Hosts," *Philosophical Transactions of the Royal Society, B* 291 (1981): 451–524.

11. D. J. Bradley, "Human Pest and Disease Problems: Contrasts Between Developing and Developed Countries, in *Origins of Pest, Parasite, Disease and Weed Problems,* ed. J. M. Cherrett and G. R. Sagar (Oxford: Blackwell Scientific Publications, 1977): 329–46.

12. See, for example, M. Burnet and D. O. White, *Natural History of Infectious Disease* (Cambridge: Cambridge University Press, 1972): 82.

13. H. D. Crofton, "A Quantitative Approach to Parasitism," *Parasitology* 63 (1971): 179–93.

14. ———, "A Model of Host-Parasite Relationships," *Parasitology* 63 (1971): 343–64.

15. R. M. May, "Host-Parasitoid Systems in Patch Environments: A Phenomenological Model," *Journal of Animal Ecology* 47 (1978): 833–44.

16. D. J. Bradley and R. M. May, "Consequences of Helminth Aggregation for the Dynamics of Schistosomiasis," *Transactions of the Royal Society of Tropical Medicine and Hygiene* 72 (1978): 262–73.

17. May and Anderson, "Population Biology. II" (See note 5).

18. Anderson and May, "Host-Parasite Interactions" (See note 8).

19. ———, "Population Biology. I" (See note 4).

20. W. G. Smillie, "Control of Hookworm Disease in South Alabama," *Southern Medical Journal* 17 (1924): 494–99.

21. J. A. Walsh and K. S. Warren, "Selective Primary Health Care: An Interim Strategy for Disease Control in Developing Countries," *New England Journal of Medicine* 302 (1979): 967–74; and subsequent "Correspondence," 302 (1979): 758–59.

22. K. S. Warren, "The Control of Helminths: Non-Replicating Infectious Agents of Man," *Annual Review of Public Health* (1981) 101–15.

23. R. H. Elderkin, D. P. Berkowitz, F. A. Farris, et al., "On the Steady State of an Age-Dependent Model for Malaria," in *Nonlinear Systems and Applications*, ed. V. Lakshmikantham (New York: Academic Press, 1977): 491–512.

24. K. Dietz, "Models for Vector-Borne Parasitic Diseases," in *Vito Volterra Symposium on Mathematical Models in Biology: Lecture Notes in Biomathematics*, vol. 39, ed. C. Barigozzi (Berlin: Springer Verlag, 1980): 264–77.

25. J. L. Aron and R. M. May, "The Population Dynamics of Malaria," in *Population Dynamics of Infectious Diseases*, ed. R. M. Anderson (London: Chapman and Hall, 1981).

26. Dietz, "Arbovirus Diseases" (See note 2).

27. R. M. Anderson, "Population Ecology of Infectious Disease Agents," in *Theoretical Ecology: Principles and Applications*, 2nd ed., ed. R. M. May (Oxford: Blackwell Scientific Publications; Amherst, Massachusetts: Sinauer, 1981): 318–55.

28. Anderson and May, "Epidemiology of Infectious Diseases" (See note 6).

29. ———, "Microparasites and Invertebrate Hosts" (See note 10).

30. W. O. Kermack and A. G. McKendrick, "A Contribution to the Mathematical Theory of Epidemics," *Proceedings of the Royal Society, A* 115 (1927): 700–21.

31. May and Anderson, "Population Biology. II" (See note 5).

32. See Bailey, *Theory of Diseases* (See note 1).

33. See J. A. Yorke, H. W. Hethcote, and A. Nold, "Dynamics and Control of the Transmission of Gonorrhea," *Journal of Sexually Transmitted Diseases* 5 (1978): 51–56.

34. Dietz, "Arbovirus Diseases" (See note 2).

35. Anderson, "Infectious Disease Agents" (See note 27).

36. Anderson and May, "Epidemiology of Infectious Diseases" (See note 6).

37. Much of this discussion is implicit in Bradley, "Human Pest and Disease Problems" (See note 11).

38. M. Greenwood, A. Bradford Hill, W. W. C. Topley, et al., *Experimental Epidemiology*, Special Report Series No. 209, Medical Research Council (London: His Majesty's Stationery Office, 1936).

39. M. Greenwood and W. W. C. Topley, "A Further Contribution to the Experimental Study of Epidemiology," *Journal of Hygiene* 26 (1925): 45–110.

40. F. Fenner, "The Epizootic Behavior of Mouse Pox (Infectious Ectromelia)," *British Journal of Experimental Pathology* 29 (1948): 69–91.

41. ———, "Mouse Pox (Infectious Ectromelia of Mice): A Review," *Journal of Immunology* 63 (1949): 341–73.

42. Anderson and May, "Population Biology. I" (See note 4).

43. Ibid.

44. Greenwood, Hill, Topley, et al., *Experimental Epidemiology* (See note 38).

45. Anderson and May, "Population Biology. I" (See note 4).

46. Ibid.

47. R. M. Anderson and R. M. May, "Infectious Diseases and Population Cycles of Forest Insects," *Science* 210 (1980): 658–61.

48. ———, "Microparasites and Invertebrate Hosts" (See note 10).

49. J. Berger, "A Mathematical Model for the Spatial Propagation of Rabies in Fox Populations," in *Mathematical Models in Medicine: Lecture Notes in Biomathematics*, vol. II, ed. J. Berger, W. Buhlen, R. Repges, et al., (Berlin: Springer Verlag, 1976): 75–88.

50. R. M. Anderson, H. Jackson, R. M. May, et al., "The Population Dynamics of Fox Rabies in Europe," *Nature* 289 (1981): 765–71.

51. Dietz, "Arbovirus Diseases" (See note 2).

52. ———, "Incidence of Infectious Diseases" (See note 3).

53. Anderson and May, "Epidemiology of Infectious Diseases" (See note 6).

54. J. A. Yorke, N. Nathanson, G. Pianigiani, et al., "Seasonality and the Requirements for Perpetuation and Eradication of Viruses in Populations," *American Journal of Epidemiology* 109 (1979): 103–23.

55. Anderson and May, "Epidemiology of Infectious Diseases" and references therein (See note 6).

56. G. Macdonald, "The Dynamics of Helminth Infections, with Special Reference to Schistosomes," *Transactions of the Royal Society of Tropical Medicine and Hygiene* 59 (1965): 489–506.

57. ———, *Dynamics of Tropical Disease*, Collected Papers, ed. L. J. Bruce-Chwatt and V. J. Glanville (Oxford: Oxford University Press, 1973).

58. J. E. Cohen, "Schistosomiasis: A Human Host-Parasite System," in *Theoretical Ecology: Principles and Applications*, 1st ed., ed. R. M. May (Oxford: Blackwell Scientific Publications; Philadelphia: W. B. Saunders, 1976): 237–56.

59. Bradley and May, "Dynamics of Schistosomiasis" (See note 16).

60. R. M. Anderson and R. M. May, "Prevalence of Schistosome Infections within Molluscan Populations: Observed Patterns and Theoretical Predictions," *Parasitology* 79 (1979): 63–94.

61. Bradley and May, "Dynamics of Schistosomiasis" (See note 16).

62. Warren, "Control of Helminths" (See note 22).

63. Walsh and Warren, "Strategy for Disease Control" (See note 21).

64. Anderson, "Infectious Disease Agents" (See note 27).

65. Smillie, "Control of Hookworm Disease" (See note 20).

66. Anderson, "Infectious Disease Agents" (See note 27).

67. G. Macdonald, *The Epidemiology and Control of Malaria* (Oxford: Oxford University Press, 1957).

68. L. Molineaux, G. R. Shidrawi, J. L. Clarke, et al., "Assessment of Insecticidal Impact on the Malaria Mosquito's Vectorial Capacity, from Data on the Man-Biting Rate and Age-Composition," *Bulletin of the World Health Organization* 57 (1979): 265–74.

69. See, for example, K. Dietz, L. Molineaux, and A. Thomas, "A Malaria Model Tested in the African Savannah," *Bulletin of the World Health Organization* 50 (1974): 347–57.

70. Aron and May, "Dynamics of Malaria" (See note 25).

71. Ibid.

THE ROLE OF MEMBRANE STUDIES IN PARASITOLOGY

Victor Nussenzweig and
Ruth S. Nussenzweig

We will begin with a statement of the obvious: during their evolution the surface membranes of protozoan parasites must accumulate the information necessary for survival in the host. Because their membranes contain immunogenic molecules the surfaces of parasites face the formidable challenge of the immune system. The membranes of parasites that take refuge inside cells must contain the molecules or apparatus that set in motion events leading to internalization. For immunologists interested in vaccines it is of course important to understand the mechanisms involved in these membrane-associated functions and to try to channel the host's defenses against those key molecules essential for parasite survival.

Recent developments in relation to African trypanosomiasis and malaria permit some measure of optimism about our ability to clarify the nature of these membrane-associated parasitic adaptations.

African trypanosomes spend their entire life outside cells and are therefore particularly vulnerable to the immune system. After inoculation of the parasites by the tsetse fly the number of trypanosomes increases rapidly in the blood of the host, and if left unchecked the infection leads to its death. Antibodies are produced that kill most of the parasites, however, and the few that remain have an altered antigenic profile;

they multiply and are in turn destroyed. The disease is characterized by successive waves of parasitemia that may continue for a long time. This type of adaptation is interesting in many ways. The immune response essential for the survival of the host also contributes to the preservation of the gene pool of the parasites, as the destruction of the host precludes their survival. For the preservation of the species, therefore, the parasite does not evade the immune response, but, rather, welcomes it.

The work of George A. M. Cross and his collaborators has clearly demonstrated that a single parasite molecule is involved in this remarkable adaptation.[1-3] It is a glycoprotein with a molecular weight of approximately 60,000 that covers the entire surface of the trypanosomes and changes periodically with each wave of parasitemia. The number of possible variant glycoproteins is not known, but it may run into the hundreds. These various proteins differ widely in amino acid sequence, and thus antibodies against one frequently do not interact with any of the others.

Although the genetic basis and regulatory mechanisms underlying the antigenic variation in trypanosomes are not entirely clear, it is likely that the DNA rearrangements that occur are similar to those recently described for a growing number of other genes. Whatever the explanation at the DNA level, the fact remains that a single protein on the surface of these flagellates may account for the host-parasite relationship. The simplicity of this solution must have been a shock to many parasitologists, who, in general, tend to believe that complexity of antigens and immune mechanisms is inherent to their discipline.

We will now turn to an example of an interaction between the host immune response and a membrane antigen of an intracellular parasite, which presumably interferes with the ability of the parasite to penetrate the target cells. We refer to recent work in our laboratories on the mechanism of the immune response of rodents to sporozoites of the malaria parasite *Plasmodium berghei.*[4,5]

Malaria is caused by a protozoan parasite of the genus *Plasmodium,* and the major species that infect humans are *P. falciparum* and *P. vivax.* Malaria is a severely debilitating disease with widespread distribution, particularly in less-developed tropical countries, and despite major expenditures for its control it still remains a major public health problem.

The life cycle of the parasite is complex. It is transmitted by the bite of a mosquito whose salivary glands contain the infective stage—mature sporozoites. When these reach the blood of the host they circulate for a short time and then penetrate the primary target cells, presumably the hepatocytes. The sporozoites differentiate and multiply rapidly in the cells, originating thousands of merozoites infective for erythrocytes. The red cells are ruptured, new cycles of reproduction take place, and the released merozoites perpetuate the blood infection. Some time after the initial infection the erythrocytes may contain the sexual forms of the parasite, which are infective for mosquitoes. The male and female gametes fuse in the stomach of the mosquito and transform into oocysts. Sporozoites develop within the oocysts, rupture them, migrate through the hemocele, and finally reach the salivary glands; during this period of migration the sporozoites become fully mature.

A. H. Cochrane and colleagues have demonstrated that the mature sporozoites of *P. berghei*, inactivated by exposure to X rays, can, against challenge, immunize mice with active sporozoites. The antigenic moiety is present in mature sporozoites, but not in the immature forms found inside the oocysts of the midgut of the mosquito. Moreover immunity is stage-specific, that is, the sporozoite-vaccinated mice are susceptible to infection with blood stages of the parasite.[6]

Vaccination of mice with *P. berghei* sporozoites is achieved by the intravenous route in the absence of adjuvants, as well as through the repeated bite of X-irradiated infected mosquitoes. Using a similar methodology for vaccination, encouraging results have also been obtained in rhesus monkeys injected with irradiated *P. cynomolgi* and *P. knowlesi* sporozoites and in human volunteers immunized by the bite of irradiated mosquitoes infected with *P. falciparum* or *P. vivax*. In more recent experiments protection has been effective against different geographic isolates of *P. falciparum*, which agrees with observations in the rodent system.

Experiments designed to study the role of circulating

antibodies have furthered our understanding of the mechanism of protection. Two observations are of particular importance: first, several experiments have demonstrated that the serum of protected rodents and monkeys contains antibodies capable of neutralizing active sporozoites in vitro; second, it has been found that incubation of immune serum with sporozoites results in accumulation of a precipitate on their membrane, which lengthens progressively at the posterior end of the parasite—circumsporozoite (CSP) reaction. These findings suggest that antibodies against a surface antigen might be involved in the protective effects of serum.

This theory is supported by several other observations. The intensity of the CSP reactions, for example, correlates well with protective immunity, particularly in primates. Adult humans living in endemic areas are known to be more resistant to infection, and their serum has significantly higher CSP titers than that of susceptible children.[7] Furthermore the characteristics of the CSP reaction and protective immunity are similar; both are stage- and species-specific, but not strain-specific.

Using monoclonal antibodies, a protective antigen from sporozoites of *P. berghei* has recently been identified. The antigen, Pb-44, is a protein uniformly distributed on the surface of the parasite with a molecular weight of 44,000 and a relatively low isoelectric point—$pI = 3.9$ in nondenaturing conditions. Cross-linking of Pb-44 by the monoclonal antibodies induces the CSP reaction. Pb-44 is absent from the surface of most oocyst sporozoites, and disappears soon again after the sporozoites penetrate the hepatocytes of the mammalian host. That is, Pb-44 is a differentiation antigen most likely involved with a sporozoite-specific function such as penetration into cells. Purified monoclonal antibodies to Pb-44—as well as the monovalent fragments of the antibodies—injected intravenously into mice afford complete protection against infection by viable sporozoites, but not by merozoites. The dose of antibodies necessary for protection depends on the number of parasites used for challenge. For example, 10 μg of antibody protect a mouse against challenge with 10^3 sporozoites, but 100 to 300 μg are necessary to protect against 10^4 parasites. To our knowledge this is the first time a protective antigen from a mammalian protozoa parasite has been identified.[8,9]

A similar study is underway to identify protective antigens on the surface of sporozoites of other species of malaria, including those that infect humans. Monoclonal antibodies against *P. knowlesi* sporozoites have recently been shown, in vitro, to neutralize their infectivity for monkeys. These antibodies react with a surface antigen of the sporozoites, induce a CSP reaction, and are also strictly stage-specific. Although the antigen has not yet been fully characterized it may be homologous to Pb-44.[10]

Work on human malaria parasites is progressing more slowly due to difficulties in obtaining sufficient numbers of sporozoites from *P. falciparum* or *P. vivax* and in developing convenient primate models to test the functional activity of hybridomas. Although Pb-44 can be isolated from sporozoites in a highly purified condition by a combination of affinity chromatography and isoelectric focusing, the amounts obtained are severly limited by the practical difficulty of accumulating large numbers of sporozoites. With the use of modern recombinant DNA technology, however, it should in principle be possible to identify and clone the genes that code for the protective antigens, produce sufficient amounts, and attempt to immunize animals against malaria infection. The results of experiments with metabolic labeling of sporozoites in vitro by incubation with ^{35}S-methionine indicate that Pb-44 is one of the major proteins synthesized by the parasite.[11] Thus the identification and isolation of the mRNA and of the gene for Pb-44 could be attempted.

We should point out that, for the purpose of developing vaccines, hybridoma technology is being used to study the functional properties of membrane antigens of the asexual and sexual blood stages of the malaria parasite. R. R. Freeman and colleagues, for example, have obtained a series of monoclonal antibodies directed against blood stages of *P. yoelii* and infected erythrocyte membrane.[12] Only those reacting exclusively with the merozoite stage of the parasite are protective. After passive transfer of antibodies the normally fulminating *P. yoelii* infection is retarded and follows a course more typical of a mild strain.

The antigens or antigenic determinants against which these monoclonal antibodies are directed have not yet been identified, but they do reside on the membrane. L. H. Perrin and coworkers have recently succeeded in obtaining monoclonals that interfere with the growth of the erythrocytic forms of *P. falciparum* in vitro.[13] Thus it is fair to say that the information resulting from the application to parasites of new technology to study surface membranes raises new hopes for the identification of protective antigens and the development of vaccines.

NOTES

1. G. A. M. Cross, "Identification, Purification and Properties of Clone-Specific Glycoprotein Antigens Constituting the Surface Coat of *Trypanosoma brucei*," *Parasitology* 71 (1975): 393–417.

2. _____, "Antigenic Variation in Trypanosomes," *Proceedings of the Royal Society B* 202 (1978): 55–72.

3. J. H. J. Hoeijmakers, A. C. C. Frasch, A. Bernards, et al., "Novel Expression-Linked Copies of the Genes for Variant Surface Antigens in Trypanosomes," *Nature* 284 (1980): 78–80.

4. N. Yoshida, R. S. Nussenzweig, P. Potocnjak, et al., "Hybridoma Produces Protective Antibodies Directed Against the Sporozoite Stage of Malaria Parasite," *Science* 207 (1980): 71–73.

5. P. Potocnjak, N. Yoshida, R. S. Nussenzweig, et al., "Monovalent Fragments (Fab) of Monoclonal Antibodies to a Sporozoite Surface Antigen (Pb-44) Protect Mice Against Malarial Infection," *Journal of Experimental Medicine* 151 (1980): 1504–13.

6. A. H. Cochrane, R. S. Nussenzweig, and E. H. Nardin, "Immunization Against Sporozoites," in *Malaria*, vol. 3, ed. J. P. Kreier (New York: Academic Press, 1980): 163–202.

7. E. H. Nardin, R. S. Nussenzweig, J. A. McGregor, et al., "Antibodies to Sporozoites: Their Frequent Occurrence in Individuals Living in an Area of Hyperendemic Malaria," *Science* 206 (1979): 597–99.

8. Yoshida, Nussenzweig, Potocnjak, et al., "Protective Antibodies" (See note 4).

9. Potocnjak, Yoshida, Nussenzweig, et al., "Monovalent Fragments" (See note 5).

10. A. H. Cochrane, R. Gwadz, V. Nussenzweig, et al., manuscript in preparation.

11. Ibid.

12. R. R. Freeman, A. J. Trejdosiewicz, and G. A. M. Cross, "Protective Monoclonal Antibodies Recognising Stage-Specific Merozoite Antigens of a Rodent Malaria Parasite," *Nature* 284 (1980): 366–68.

13. L. H. Perrin, E. Ramirez, P. H. Lambert, et al., "Inhibition of *P. falciparum* Growth in Human Erythrocytes by Monoclonal Antibodies," *Nature* 289 (1981): 301–03.

THE STUDY OF PARASITISM AS A BRANCH OF BIOLOGY

William Trager

Parasitism involves an intimate association between two different kinds of organisms: one of these, the host, provides food and shelter for the other, the parasite. The host may or may not be injured by the parasite, which it may quickly expel or destroy or may harbor for many years. Because the parasite cannot exist alone in nature, it is not to its advantage to destroy the host, or at least not until it or its progeny are ready to move to another. Some hosts benefit from their parasites, and some are even dependent on specific parasites—a special type of association called mutualism. Such mutualistic associations may have been the origin of chloroplasts and mitochondria, and so may be the basis of most eukaryotic cells.

Throughout the living world, from prokaryotes to man, parasitic associations are extremely common. It can safely be said that, except for the viruses, all organisms have parasites; furthermore, parasitic organisms are found in all main toxonomic groups. To study parasites as organisms in their own right is relatively simple and straightforward, but to study the interrelations between the parasite and its host, that is, to study parasitism, one must draw on all disciplines of biology from ecology to biophysics.

In the past most workers in animal parasitology first described and classified the parasites and then dealt with their life cycles. More recently the metabolism of the parasite has been investigated, and although such studies provide essential background they do not attack the basic problem of the parasitic relationship. It is the study of the cell biology and physiology of host-parasite relationships that will constitute the main body of future work. It is clear that the immunology of parasitic infections—dealing almost exclusively with the acquired immunity of vertebrates—is only a branch of this broad subject. In its practical as well as basic aspects, however, it is such an important branch that it is considered at length in other chapters in this book. Accordingly, I will try to provide a few brief examples drawn from aspects of host-parasite relations other than the immunological.

The complex life cycles of parasites provide biologists with a remarkable series of differentiations triggered by external factors. This is seen in the life cycles of trypanosomes— flagellated protozoa that cause sleeping sickness in man and nagana in cattle in Africa. When multiplying in the bloodstream of their vertebrate host, these organisms have a fermentative type of metabolism not mediated through cytochrome oxidase. When ingested by a tsetse fly the organisms multiply in its gut and switch to primarily cytochrome oxidase-mediated respiratory metabolism, at the same time losing their infectivity for the vertebrate host. Some of the gut trypanosomes invade the salivary glands or proboscis of the fly, and after a further multiplicative cycle they become very similar to bloodstream forms, ready to infect another vertebrate host when injected by the bite of the fly. These metabolic changes are accompanied by certain morphological changes that have been the basis of now extensive studies in the molecular biology of the trypanosomes.

Trypanosomes have a characteristic organelle, the kinetoplast, consisting of a disc-shaped or spherical portion containing much DNA (up to 20 percent of total cellular DNA) and a long tubular part with a typical mitochondrial structure. Since the life cycle changes are accompanied by changes in the extent of elaboration of mitochondrial material, and since the kinetoplast DNA (K-DNA) presumably represents a kind of mitochondrial DNA, it has been hypothesized that the cellular differentiations in the life cycle might be under the control of K-DNA. Whether this is so remains to be demonstrated; meanwhile it has become clear that K-DNA is a remarkable biological material.[1] It consists of an extensive network of catenated minicircles of differing nucleotide sequences, plus a small proportion of larger so-called maxicircles resembling the usual circular DNA of other kinds of mitochondria. When we know what messages these DNAs are responsible for we may begin to understand the cyclical development of trypanosomes. Some dyskinetoplastic trypanosomes have only a bloodstream form and cannot develop in tsetse flies: they seem to lack minicircles. If this is indeed so it will point strongly to their role in cyclical development. There is much work here for future biologists, who will at the same time contribute to our understanding of parasitism and of developmental biology.

The trypanosome cell membrane is involved in interactions with lipoproteins of the host plasma, which determine host specificity. *Trypanosoma brucei brucei* is by definition distinguished from *T.b. rhodesiense* by a single property: the former is noninfective to man although infective to domestic animals, whereas the latter is infective to man and to domestic animals and is a cause of African sleeping sickness. Human serum in vitro or in vivo quickly renders *T. b. brucei* noninfective to mice, but it has no such effect on *T. b. rhodesiense*.

The effects of human serum can be fully reproduced with purified human high-density lipoprotein (HDL).[2] Rat HDL has no effect, but it is higher in phosphatidyl inositol than human HDL. M. R. Rifkin has recently found that the toxicity of human HDL can be markedly suppressed with pure phosphatidyl inositol.* While little is known of the lipid composition or metabolism of trypanosomes, it is clear that extremely subtle variations may mean the difference between life and death; we are now in a position to try to find out what these variations are. With the development of liposome technology we can imagine introducing a lipid factor that would render the membrane of *T. b. rhodesiense* susceptible to human HDL, just as the membrane of *T. b. brucei* is. Here we have the basis for a study of a whole new aspect of the parasite-host interface: the interaction

---

* M. R. Rifkin, 1980: personal communication.

between lipids of the parasite plasma membrane and the lipids and lipoproteins of the host.

The complexity of host-parasite interaction is probably best seen in intracellular parasitism. Whereas only a few helminths such as *Trichinella spiralis* develop intracellularly, whole groups of protozoa do so. Unlike viruses, the obligate intracellular protozoa have their own protein-synthesizing machinery, and it is not clear why they can develop only inside other living cells. The malaria parasites have been studied most extensively,[3] and there is evidence that they require exogenous sources of cofactors, such as coenzyme A, ordinarily present only within cells.

The recent successful propagation of human malaria *Plasmodium falciparum* in human erythrocytes in vitro has opened up a new field. Not only is the human erythrocyte the best-known cell, but numerous variants of it exist in nature. The geographic distribution of some of the most prevalent of these variants shows a strong correlation with the distribution of falciparum malaria. All indications are that individuals with these variant cells are relatively resistant to the often lethal effects of falciparum malaria. The selective pressure of this disease has selected for these variants.[4]

Best studied of the variants have been sickle hemoglobin (HbS) and glucose-6-phosphate dehydrogenase (G6PD) deficiency. The situation with HbS is completely clear: with regard to resistance to lethal malaria the heterozygotes (AS) have a sufficient advantage over children with only normal hemoglobin (AA) to offset the lethal effects of the homozygous (SS) condition. (SS individuals usually die before reaching reproductive age.) M. J. Friedman and coworkers recently demonstrated the physiological basis for this resistance.[5] When AS cells are under a relatively low oxygen tension, in the area of 5 percent—a condition that occurs normally in many parts of the body—they become more permeable and lose potassium, thus rendering them unsuitable for the development of malaria parasites. The situation with regard to G6PD deficiency is more complex. Here again, however, Friedman has shown by in vitro experiments that prooxidant conditions, as a 25 percent oxygen atmosphere, or addition of menadione or riboflavin to the medium, render G6PD-deficient cells unsuitable for *P. fal-ciparum*, whereas normal cells are unaffected.[6] It is of interest that for a human red cell to become unsuitable for *P. falciparum* it must sacrifice something and become a less fit cell. Thus HbS is subject to paracrystalline formation under physiological low oxygen tension, with very deleterious results in the homozygous condition. A number of different loci may be responsible for G6PD deficiency, which leads to sensitivity of the cells to hemolysis by certain foods such as fava beans, and by certain drugs such as primaquine. This emphasizes the remarkably good adaptation of the parasite to the normal human red cell.

Evidence has been obtained very recently that the red cells of people with ovalocytosis are less susceptible than normal cells to invasion by *P. falciparum* merozoites.* Ovalocytosis is a red cell variant common in New Guinea that affects the cell membrane. Further work might provide a key to membrane factors involved in attachment to and penetration of the red cell by malarial merozoites. Only for *P. knowlesi* in rhesus monkeys is there any direct evidence on this subject; merozoites attach to but do not enter Duffy-negative human erythrocytes. Full interpretation of this significant finding is made difficult by the fact that knowlesi merozoites readily enter certain kinds of Duffy-negative monkey erythrocytes. Nevertheless this work did lead to experiments showing that Duffy-negative humans are refractory to the human malaria *P. vivax*.[7] This explains the absence of *P. vivax* in West Africa, where 99 percent of the population is Duffy-negative.

In the immediate future there will be much further research on how intracellular parasites attach to and enter their host cells. The morphological events have already been described in some detail by M. Aikawa and coworkers.[8] For organisms such as malarial merozoites it is clear that the apical end of the merozoite induces invagination of the red cell membrane to receive the parasite into a parasitophorous membrane. Very quickly the nature of the invaginated membrane alters— the distribution of intramembranous particles changes and at least two membrane enzymes reverse their polarity. How all this

---

* C. Kidson, 1980: personal communication.

occurs remains a fascinating question in cell biology. In malaria the parasitophorous membrane becomes an important functional part of the parasite, being closely apposed to its plasma membrane throughout the parasite's growing stage. In *Babesia,* a related intraerythrocytic parasite, the parasitophorous membrane disappears soon after its formation, so the parasite lies directly in the cytoplasm of the erythrocyte. These two situations must create considerable differences in the metabolic interchanges between parasite and host cell; a comparative study would be most interesting. The potential practical value of such a study is indicated by the great differences in drug susceptibility between *Plasmodium* and *Babesia.*

Recent work with malaria parasites indicates that they not only take over the parasitophorous membrane, but indeed modify the whole host cell. The presence of malarial antigen on the surface of the infected red cell has been shown for several species. It is most striking in *P. falciparum,* where structures visible by electron microscopy appear as protrusions or knobs on the surface of the host erythrocyte. The knobs appear at the trophozoite stage when the infected red cells adhere to capillary endothelium and become sequestered in certain organs, and there is evidence that they are responsible for this adherence.[9]

Modification of the host cell probably finds its most striking expression in cell hypertrophy, seen with many different kinds of intracellular parasitic associations. This effect reaches an extreme in certain microsporidian parasites of fish, where great tumorous formations, called xenomas, arise. When associations of this type can be brought into the laboratory, as perhaps in a suitable fish tissue culture, they cannot fail to lead to significant discoveries in both cell biology and the biology of parasitism.

This brings me to one other point: the importance to the future of parasitology not only of investigators who will use material already available to study the biology of parasitism, but of those who will bring into the laboratory and make available new kinds of experimental material. In this connection the further development of in vitro culture methods must continue to have a high priority, for they permit dissection of the parasitic association and the study of its component parts and of their interactions with each other.

Finally it is important to state that parasitism and symbiosis are such major biological phenomena that their study in a modern context should be a part of every biologist's training, beginning at the undergraduate level. If granting agencies and foundations wish to support continued research on parasitism, one of the most valuable contributions would be to help establish positions in biology departments for the teaching of undergraduate courses in parasitology that embody new approaches in this field.

NOTES

1. P. Borst and J. H. J. Hoeijmakers, "Kinetoplast DNA," *Plasmid* 2 (1979): 20–40.

2. M. R. Rifkin, "Identification of the Trypanocidal Factor in Normal Human Serum: High Density Lipoprotein," *Proceedings of the National Academy of Sciences* 75 (1978): 3450–54.

3. I. W. Sherman, "Biochemistry of Plasmodium (Malarial Parasites)," *Microbiology Review* 43 (1979): 453–95.

4. L. Luzzatto, "Genetics of Red Cells and Susceptibility to Malaria," *Blood* 54 (1979): 961–78.

5. M. J. Friedman, E. F. Roth, R. L. Nagel, et al., "*Plasmodium falciparum*. Physiological Interaction Between the Malaria Parasite and the Sickle Cell," *Experimental Parasitology* 47 (1979): 73–80.

6. M. J. Friedman, "Oxidant Damage Mediates Variant Red Cell Resistance to Malaria," *Nature* 280 (1979): 245–47.

7. L. H. Miller, M. H. McGinnis, P. V. Holland, et al., "The Duffy Blood Group Phenotype in American Blacks Infected with *Plasmodium vivax* in Vietnam," *American Journal of Tropical Medicine and Hygiene* 27 (1978): 1069–72.

8. M. Aikawa, L. H. Miller, J. Johnson, et al., "Erythrocyte Entry by Malarial Parasite. A Moving Junction Between Erythrocyte and Parasite," *Journal of Cell Biology* 77 (1978): 72–82.

9. W. Trager, "Cultivation of Malaria Parasites *in Vitro*: Its Application to Chemotherapy, Immunology and the Study of Parasite-Host Interactions," in *The Host-Invader Interplay*, ed. H. Van den Bossche (New York: Elsevier/North-Holland, 1980): 537–48.

DISCUSSANTS:

Larry Simpson

I would reemphasize William Trager's conclusion that advances in parasitology rely heavily on the development and dissemination of in vitro culture methods. The continuous culture of falciparum malaria in red blood cells is one of the strongest examples of this truism. Another is the development of an in vitro culture system for the bloodstream forms of the African pathogenic trypanosomes by H. Hirumi[1] and G. C. Hill[2] and their coworkers. The bloodstream trypomastigotes are cultured at 37° C in the presence of a feeder layer of mammalian tissue culture cells. The appearance of variable surface antigens in the bloodstream trypomastigotes in culture and the possible attenuation of virulence in cultured trypomastigotes make this system attractive for studies of the basic mechanisms involved in variable antigen modulation and virulence. In addition, Hirumi has succeeded in reproducing the entire life cycle of *Trypanosome brucei* in vitro. In this regard our group has developed a culture system that allows fairly synchronous morphological differentiation of cultured bloodstream trypomastigotes into procyclic trypomastigotes.[3] This system is useful for the study of the biosynthesis of mitochondrial enzymes that occurs during this time.

The study of the unusual mitochondrial DNA in the trypanosomes mentioned by Trager is my main interest and I would like to elaborate somewhat on that theme.

The kinetoplast DNA, as the unusual mitochondrial DNA of the kinetoplast protozoa is called, consists of thousands of catenated minicircles ranging in size from 0.9 Kb to 2.5 Kb, and twenty to fifty larger maxicircles ranging in size from 20 to 40 Kb in different species.[4-6] The minicircles consist of several to many different sequence classes, all sharing a small constant region.[7,8]

The genetic function, if any, of the minicircle is still unknown; the maxicircle appears to represent the homologue of the mitochondrial DNA in other organisms. It codes for the small mitochondrial ribosomal RNAs (9 and 12s RNAs) and for several other polyadenylated transcripts that are presumptive mRNAs.[9,10] Detailed restriction maps are available for maxicircle DNA from several *T. brucei* strains[11,12] and from *Leishmania tarentolae*.[13] The ribosomal sequences are highly conserved between species,[14] and some conservation of nonribosomal maxicircle gene sequences and gene arrangements has also been found.* Portions of the maxicircle are extremely rich in adenine and thymine and seem to vary extensively among species and even between strains or stocks of the same species.[15] It is believed, but not yet proven, that these represent nontranscribed spacer regions of the maxicircle. Several maxicircle fragments and several minicircles have been cloned in plasmids and phages in *E. coli* and can be obtained in large quantities in homogenous form.[16,17] In fact the application of recombinant DNA technology to the kinetoplast DNA problem has resulted in a molecular dissection of this unusual genetic system that could not otherwise have been performed.

In contrast to the maxicircle DNA, kinetoplast minicircle DNA has a rapid rate of sequence evolution in nature. Minicircle sequence differences can even be observed between different stocks or strains of the same species. C. E. Morel and I have used these sequence differences to classify different stocks of the pathogenic trypanosome, *T. cruzi*, into groups or "schizodemes" that exhibit identical minicircle restriction profiles in acrylamide gels.[18] All stocks from zymodeme group C belonged to a single schizodeme, whereas stocks from zymodeme groups A and B were classified into several schizodemes. Control experiments have shown that the minicircle restriction

---

* L. Simpson, A. M. Simpson, J. Borst, et al.: unpublished observations.

profile changes slowly enough in the laboratory so as not to preclude gel analysis after establishment of a primary hemoculture of trypanosomes. The method is rapid and simple and requires only a small volume of an established trypanosome hemoculture. We believe that schizodeme analysis will extend and supplement such standard methods of classification as zymodeme analysis.

Another parasitological problem that addresses a basic biological question is the genetic control of the variable surface antigen switch that occurs in the bloodstream phase of the life cycle of the African pathogenic trypanosomes. Again, the use of recombinant DNA techniques has led to the conclusion that expressed genes coding for surface antigens actually change their position in the chromosome.[19-21] This adds to the growing list of differentiation-associated genetic rearrangements that have recently become known in other biological systems. In the case of the trypanosome, a resolution of this genetic rearrangement in molecular terms may not only increase our knowledge of this basic biological phenomenon, but may aid in the development of a vaccine against the trypanosome.

I shall conclude by reemphasizing Trager's conclusion that the study of parasite-host systems depends to a great extent on in vitro culture systems, and by emphasizing that the use of the techniques of recombinant DNA should allow molecular dissections of the genetic systems of the parasite and host that are involved in the establishment of the parasitic way of life. Studies of the basic physiological and genetic mechanisms involved in parasitism should certainly lead to additional basic discoveries in cell and molecular biology.

Notes

1. H. Hirumi, J. Doyle, and K. Hirumi, "African Trypanosomes: Cultivation of Animal-Infective *Trypanosoma brucei* in Vitro," *Science* 195 (1977): 992–94.

2. G. C. Hill, S. Shriner, B. Caughey, et al., "Growth of Infective Forms of *Trypanosoma rhodesiense* in Vitro: The Causative Agent of African Trypanosomiasis, *Science* 201 (1978): 763–65.

3. L. Simpson, A. Simpson, G. Kidane, et al., "The Kinetoplast DNA of the Hemoflagellate Protozoa," *American Journal of Tropical Medicine and Hygiene* 29 (suppl.) (1970): 1053–63.

4. L. Simpson, "The Kinetoplast of the Hemoflagellates," *International Review of Cytology* 32 (1972): 130–207.

5. P. Englund, "Kinetoplast DNA," in *Biochemistry and Physiology of Protozoa*, vol. 4, ed. M. Levandowsky and S. Hutner (New York: Academic Press, forthcoming).

6. P. Borst and J. H. J. Hoeijmakers, "Kinetoplast DNA," *Plasmid* 2 (1979): 20–40.

7. K. Chen and J. Donelson, "The Sequences of Two Kinetoplast DNA Minicircles from *Trypanosoma brucei*," *Proceedings of the National Academy of Sciences* 77 (1980): 2445–49.

8. Simpson, Simpson, Kidane, et al., "Kinetoplast DNA" (see note 3).

9. L. Simpson and A. Simpson, "Kinetoplast RNA of *Leishmania tarentolae*," *Cell* 14 (1978): 169–78.

10. Simpson, Simpson, Kidane, et al., "Kinetoplast DNA" (see note 3).

11. P. Borst and F. Fase-Fowler, "The Maxicircle of *Trypanosoma brucei* Kinetoplast DNA," *Biochimica et Biophysica Acta* 565 (1979): 1–12.

12. K. Stuart, "Kinetoplast DNA of *Trypanosoma brucei*: Physical Map of the Maxicircle," *Plasmid* 2 (1979): 520–28.

13. H. Masuda, L. Simpson, H. Rosenblatt, et al., "Restriction Map, Partial Cloning and Localization of 9s and 12s RNA Genes on the Maxicircle Component of the Kinetoplast DNA of *Leishmania tarentolae*" *Gene* 6 (1979): 51–73.

14. Simpson and Simpson, "Kinetoplast RNA" (see note 9).

15. P. Borst, F. Fase-Fowler, J. H. J. Hoeijmakers, et al., "Variations in Maxi-Circle and Mini-Circle Sequences in Kinetoplast DNAs from Different *Trypanosoma brucei* Strains," *Biochimica et Biophysica Acta* 610 (1980): 197–210.

16. Masuda, Simpson, Rosenblatt, et al., "Restriction Map" (see note 13).

17. Simpson, "Kinetoplast of Hemoflagellates" (see note 4).

18. C. E. Morel, C. Chiari, E. Camargo, et al., "Strains and Clones of *Trypanosoma cruzi* Can Be Characterized by Restriction Endonuclease Fingerprinting of Kinetoplast DNA Minicircles," *Proceedings of the National Academy of Sciences* 77 (1980): 6810–14.

19. J. H. J. Hoeijmakers, A. Frasch, A. Bernards, et al., "Novel Expression-Linked Copies of the Genes for Variant Surface Antigens in Trypanosomes," *Nature* 284 (1980): 78–80.

20. R. Williams, J. Young, and P. Majaiwa, "Genomic Rearrangements Correlated with Antigenic Variation in *Trypanosoma brucei*," *Nature* 282 (1979): 847–49.

21. J. H. J. Hoeijmakers, P. Borst, J. Van den Burg, et al., "The Isolation of Plasmids Containing DNA Complementary to Messenger RNA for Variant Surface Glycoproteins of *Trypanosoma brucei*," *Gene* 8 (1980): 391–417.

George C. Hill

The three points I want to discuss deal with the biochemical differences that occur during the life cycle of parasites; some of the mechanisms that develop may in fact be more complex than we initially anticipated.

For example, we know that antigenic variation occurs, but, as Joshua Lederberg has pointed out, "this may or may not be helpful to us in the development of a vaccine." The recent clear evidence for gene rearrangement makes this phenomenon even more complex than may have been expected.

A second example in trypanosomes of biochemical complexity is important: aerobic glycolysis occurs in them, and an α-glycerophosphate oxidase (α-GP-oxidase) is present that is insensitive to cyanide, but is inhibited by salicylhydroxamic acid (SHAM) (Figure 1). Initially it was thought that by inhibiting this oxidase in the bloodstream trypomastigotes, one could completely inhibit the respiration of trypanosomes. As it turns out, however, if the aerobic glycolysis of the trypanosomes is inhibited, the organisms use an anaerobic glycolytic scheme, with the end products being glycerol and pyruvate. The identity of this anaerobic pathway is unknown. Thus, while it is possible to study these organisms and identify unique enzymes, in order to

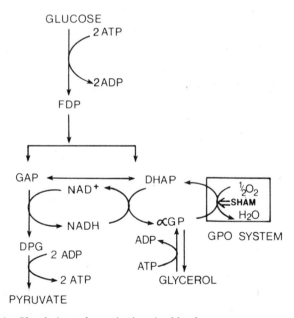

Figure 1. Glycolysis and respiration in bloodstream trypanosomes, FDP = fructose-1, 6-diphosphate; GAP = glyceraldehyde-3-phosphate; αGP = α-glycerophosphate; DPG = 1, 3-diphosphoglycerate; DHAP = dihydroxyacetonephosphate; and GPO = α-glycerophosphate oxidase.

take advantage of these findings a thorough characterization of
the biochemical system being investigated is often required.

The final point I would like to make is the importance of
trypanosomes and several other parasites as models for the
biochemical changes that occur during transformation and differentiation.

William Trager and Larry Simpson alluded to the fact that
one can now observe in vitro the transformation of organisms
from bloodstream trypomastigotes to procyclic trypomastigotes; what is puzzling, however, is that, during transformation, in two or three days the organisms differentiate morphologically into procyclic trypomastigotes. As can be seen in
Figure 2, however, in established procyclic trypomastigotes it
takes three to four weeks for a complete cytochrome system
with cytochrome oxidase to develop. In addition, the procyclic
trypomastigotes retain the α-GP-oxidase, and thus have a
branched electron transport system with two oxidases (Figure
3). After developing as metacyclic trypomastigotes in the salivary glands of the tsetse fly, on inoculation into vertebrates the
bloodstream trypomastigotes retain the α-GP-oxidase and lose
the cytochrome system.

Thus the organisms are extremely complex. Regardless of
the type of biochemical investigations we do, we continue to
find more complex systems requiring further investigations.

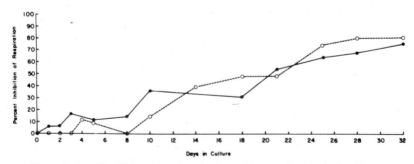

Figure 2. Cyanide (CN-) and antimycin A-sensitive respiration during *Trypanosoma
brucei* LUMP 1026 tranformation. O−−−O = CN- sensitivity in the absence of SHAM;
●−−−● = antimycin A sensitivity in the absence of SHAM.

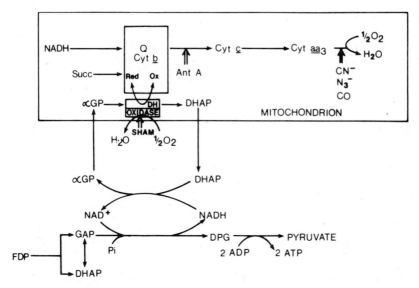

Figure 3. The proposed arrangement of respiratory systems in the procyclic trypomastigotes of *Trypanosoma brucei*. The abbreviations used here are the same as in Figure 1. Succ = succinate; Red and Ox = unidentified components that can transfer electrons between the GPO and the cytochrome system.

THE INTERFACE OF MODERN SCIENTIFIC RESEARCH AND PARASITOLOGY

Paul R. Gross

Let me begin with a demurrer concerning my assigned title. I am supposed to discuss "The Interface of Modern Scientific Research and Parasitology," and I could perhaps have tried to do so, but I do not believe there is a sufficient discontinuity between parasitology and "modern scientific research" to recognize an interface, at least of the ordinary kind, between them. There is, however, an interfacial character of parasitology *itself* that convinced some of us at the Marine Biological Laboratory (MBL) in Woods Hole to create a course on "The Biology of Parasitism."

I have another demurrer, to do specifically with the term "modern." The phrasing of the assigned title implies there is a discontinuity between parasitology and *modern* scientific research, and hence no discontinuity between parasitology and, I suppose, *immodern**** scientific research. I do not accept that implication.

Demurrers aside, I cannot tell this group what parasitology is all about; I am not qualified to do so, while many of you are. I can give my impressions of its relevance, however, and that may be interesting because I am *not* a parasitologist, but a develop-

---

* I appropriate that charming word from Mark Harris, who used it in a novel that has relevance to some questions about undergraduate education that have come up for discussion at this conference: its title is *Wake Up, Stupid!*

mental biologist. I have never taken a course in parasitology, although I did study protozoology at the University of Pennsylvania, where parasitic protozoa were not in the least neglected by D. H. Wenrich, the eminent teacher of the course.

When I came to the MBL as director in midsummer of 1978, my interest in parasitology was already strong because it is an interface subject among the traditional MBL domains of biology. The period of its great growth was at a time when the phenomena of parasitic *adaptation* presented a puzzle in the context of what had just come to be generally understood about evolution. Because the puzzle was so powerful it had to be addressed, and, as that began, the astonishing *versatility* of parasites emerged. That made parasitology interesting in its own right. It was a subject then at the interface between early post-Darwinian evolutionary biology and genetics and cytology, just before their explosive growth. That is how it was taught in Woods Hole in the early 1900s, as a matter of fact.

Now parasitology is once again a frontier subject, or about to become one, for reasons that must be quite evident to the parasitologists at this conference. This is not so much because it lies at the useful edge of molecular biology, but because it lies between molecular biology and the extended reach of modern ecology. Robert May has already discussed ecology in such a context.

My point is simply that there is a potential within ecological genetics at the present time—within the reach of tools now available to ecologists who model ecosystems on the scale computer technology makes possible—to understand the rules that determine not only how parasities are distributed in nature, but how they came to be so distributed. So far I know of very little of this kind of work being done, but the potential is surely there, and that makes it all the more interesting, particularly for the MBL. It is a matter of enormous importance, not only for theoretical ecology as I expect it will develop in the next decade, but quite obviously for epidemiology, the practical side of the matter.

Finally, because parasites have some of the most bizzare and, at the same time, discrete developmental switches known among multicellular animals, and because these clearly involve alteration either in the structure or in the utilization of genes, it seems important for people in my own field to reexamine the range of research systems possible among well-studied parasites. My first thoughts on this matter occurred long before the successful laboratory culture of *Plasmodia*, but I decided it was not within *my* competence then to work on it. Now, however, it can be a quintessential Woods Hole subject for developmentalists as well as parasitologists.

That is a summary of the reasons for a *personal* interest in a resurgence of parasitology at the MBL.

As between an individual's grand ideas and the emerging academic reality, however, there can be a vast distance. The interest of our story to a group having scientific as opposed to administrative purposes—such as this one obviously has—must be that what we did at the MBL may be a prototype of what can happen elsewhere, if we care sufficiently to make it happen.

The academic arguments spread over a nine- or twelve-month year elsewhere are often compressed into the summer session in Woods Hole. That can be a problem. By the same token the intensity of information exchange and arguments that are *other* than merely political is proportionately higher at the MBL. Hence substantive issues, if they are sufficiently important, are resolved very quickly. The outcomes may not always be right, but the answers emerge.

The MBL has been offering a group of distinguished summer courses more or less continuously since the 1890s, although not always under the same names. Parasitology has not been one of these continuing teaching efforts, although there have been a number of important special offerings from time to time, and of course many important parasitologists have and do come to Woods Hole to conduct research.

The problem was to convince the MBL Corporation and trustees—the owners who are also the users of the laboratory—that some of the already limited resources should be preempted for a major new educational program in parasitology. It was conceived as the prototype of a course one would want to establish elsewhere, now that applicable technologies from the molecular side are readily available, and now that ecologists have sufficient computer power and manpower to pay attention to theoretical problems of parasitism and parasitic disease.

Convincing people of this need is a problem in places such

as the MBL because everyone observes with a jealous eye the assignment of scarce facilities to others. Every assignment of space to a course or to an investigator necessarily deprives someone else. Beyond convincing one's colleagues that something ought to be done, there were other problems: assembling the faculty; acquiring a physical plant; building laboratories suitable for the highly sophisticated teaching we were considering; providing an adequate small mammal facility; and ensuring biosafety control—regardless of our own views about the reality of perceived hazards. In order to avoid trouble we had to be scrupulous about promoting public understanding. Finally, there was the problem of obtaining adequate funds for an activity that did not fit into any governmental or institutional pigeonhole.

All these activities began in April 1979 with a planning meeting that came about due to original encouragement from Joshua Lederberg, Kenneth S. Warren, Joseph A. Cook, and a number of colleagues within the MBL family, such as Frederik B. Bang and Irwin W. Sherman. Rather quickly thereafter the things I have just listed were done, one way or another. I must add that the convincing part was at first the least successful.

I suspect that the initial convincing occurred in part because I was a new director—a sort of political honeymoon benefit—and because the new course was part of a much larger plan. In any case, after *very* full discussion, the trustees agreed with admirable directness to let us try.

I will conclude this story simply by listing a few facts about the course as it actually emerged in 1980.

• The teaching laboratory suite was completely redesigned and renovated. Although it is located in a historic and rather neglected building, which one would not expect to be an efficient headquarters, it *is* exceptionally efficient. The laboratory is as well suited to gene cloning and hybridoma culture as it is to traditional parasitological work at the microscope or operating table.

• By contemporary standards the cost of equipping it was not high, but it is a fine working laboratory in which experiments of the kind mentioned not only can be done, but, more important, can be *taught,* which is not the same thing.

• The initial major advisors for the recruitment of teaching staff were Warren and the others mentioned earlier, plus John David and Eli Chernin. Such quick and sparkling recruitment testifies to the attractions at Woods Hole, to be sure, but even more to David's persuasive powers. The course "staff," mainly John and Roberta David, eventually took over the job of planning the physical organization of the course.

• Promotion of the program was done with care, but modestly; certainly with nothing like the intensity characteristic of the older and better-known MBL summer courses such as Physiology and Embryology.

• The response was quick, broad, and encouraging. I will not report details such as the large number of applicants, but many more highly qualified people responded than we could accommodate. Sixteen places were available for the first offering in the renovated laboratory, and all were filled promptly. Final selections consisted of one premedical student, six graduate students—some of them M.D.-Ph.D. candidates—two M.D.'s, and seven Ph.D.'s. Geographical origins of this group were worldwide. Their educational level was, on the whole, typical of those in our smaller, more advanced summer courses such as Neurobiology.

• Among the faculty present throughout the course were, in addition to John David, Richard Carter, Chernin, George Cross, Carlos Gittler, Marcel Hommel, Diane McMahon, Louis Miller, and Willy Piessens. Many others contributed to the lectures and seminars.

• Lectures were given every weekday, at least, and were attended by all participants. As we had hoped, their high quality and the *exciting research opportunities* they opened up came quickly to be known throughout our summer community. Thus by the end of the season the lectures were being attended by audiences larger, and of even more diverse backgrounds, than at the start. We will have to provide bigger and better quarters for future lectures. The abandoned firehouse in which they were given, for lack of another location, has great charm, but it was not adequate for the size of the 1980 audiences.

• Techniques taught and experiments completed comprise a remarkable list. Since Warren has already presented a summary of them I shall not repeat it, but I use the word "remarkable" advisedly and in the positive sense. This is a time when formal laboratory teaching has suffered a great decline, owing in part to the expense and in part to acquiescence of department heads and deans, who accept their faculty's claim that: "Formal laboratory teaching is no longer necessary because the students learn by helping with the research in our own laboratories."

That is a handsomely self-serving statement. I have always found it hard to believe, based on the technical competence of students who have come from distinguished places to work in my own laboratory. In any event the exercises and investigations done during the 1980 MBL parasitology course were sufficient for the participants to be capable thereafter of doing such things independently and correctly at their own institutions. George C. Hill can attest to this.

The essential point is that for our student body—the kind of clientele whose attention span and enthusiasm one did not have to worry about—it was possible in a period of some ten weeks to provide sufficient practical contact with applicable techniques of molecular and cell biology to make them comfortable about going home and using them. Monoclonal antibody, recombinant DNA, and advanced microscopic techniques are apparently neither as arcane nor as difficult to teach as they are made out to be.

Investigators who use any new research technique tend to insist that it takes a lot of time and trouble to learn it. Perhaps so, but our course was and is an experiment in communicating *quickly* certain important techniques of cellular and molecular biology to a group of diversely educated people, many of whom have had no prior contact with them, yet with *important applications immediately* evident. So far the experiment appears to be successful.

My impression is that most of the participants learned the techniques quite well. But please note this: *they also learned a great deal about parasites*—and for the molecular biologists *that* was all new! The confluence of two such different streams of contemporary biology did not seem to give participants real trouble with turbulence.

Just as the lectures—and the seminars that arose out of them—were well attended and generated excitement within the immediate company, so too did they earn a certain approval from the community as a whole. I am not sure the faculty is as yet aware of it. Many outsiders looked in on the new course at some point during the summer to see how it was going. Some still marvel that we have not been attacked by Luddites for harboring "dangerous parasites."

Our conclusions about the course are in a report we sent to Cook and Warren, whose aid, and that of the two great foundations they represent, was critical to the entire undertaking. In summary, I wrote that I have participated in, or observed the first offerings of, many advanced courses, sometimes in the guise of a teacher, sometimes of a department chairman, and recently as a dean. None of these offerings was as complex, as specialized, nor as original an undertaking as the MBL "Biology of Parasitism" course. None was so quickly brought to effective and predictable function. Everyone associated with the effort, not the least among them the foundations' staff and the MBL support staff, deserves the thanks of the biological community.

I would like to conclude from this that such an undertaking can and should be assumed elsewhere. We need not dwell on the justifications for effecting a transfer of ideas and methodology between specialists in parasitic disease and molecular biologists. Whether this can be done in *undergraduate* education I do not know. My skepticism arises from the current climate of retrenchment in even the strongest biology departments. But an effort in graduate and medical education, in which the biology of parasitism is returned to the theoretical centrality it once had—because it *is* an interface subject and because it *does* promise opportunity for important discoveries—is certainly practical and eminently worthwhile. It may indeed be that trustees, or their equivalents, in institutions other than the MBL may prove to be just as understanding as were ours.

DISCUSSION

KREIER: We have been discussing what we can do to maintain parasitology as a discipline. On occasion one can undertake to save something by doing certain things, and the result is that you save the form but not the substance—the essential quality has disappeared.

I have heard a great deal of discussion here and elsewhere about the use of parasitological systems for molecular biology and other disciplines. If one thinks in these terms one may end up with something that bears as little relationship to the mission of parasitology as the use of *E. coli* for particular molecular biological studies bears to *E. coli* as a parasite.

We should therefore give some consideration to bringing the molecular biologists into parasitology, rather than making parasitology a model system on which molecular biology research can be carried out.

WARREN: The field is large enough to encompass all kinds of aspects, including a molecular biologist using parasites as a model. As Lederberg pointed out, the great advances in bacteriology and virology were made because molecular biologists used bacteria and viruses as models.

"The Biology of Parasitism" course in Woods Hole was created because Lederberg believed we needed something similar to the phage course at Cold Spring Harbor, which so stimulated molecular biology. If we encourage the molecular biologists to study parasites, this will stimulate molecular biology and there will be enormous spinoffs in terms of practical outcomes.

Simpson, who was trained as a parasitologist at the Rockefeller University and began his career there, belongs to a school that uses trypanosome kinetoplasts as a model for studying the molecular biology of DNA. In a sense that school is at the forefront of DNA research.

LEDERBERG: Kreier still has a point. That is, if we did

nothing but interest molecular biologists in using parasites as test organisms it could conceivably be another twenty years before they stumbled into the disease-related aspects of their functions. Nobody advocates that. The problem is to get it to happen, and, having started, to maintain a reasonable line of communication with more front-line-oriented kinds of work.

Klebanoff gave a very interesting presentation at several levels. One provocative aspect of it is how it again illustrates the paradigm of a natural historical observation crying out for an analysis that produces exciting results.

The observation goes back to the fact that there are obvious eosinophils in high concentration in the blood of people with parasitic infections—a little more in terms of attachment of eosinophils to parasites. But that is roughly the point at which natural history stops. For that reason attention is then focused on the eosinophil.

If one takes the other steps and studies the molecular biology of all cells, then, when we understand that theoretical structure, we will as a matter of course discover the eosinophil and its relation to parasites. That would take over, and it is not a very efficient way to proceed. But if we did not have the techniques for the fractionation of cells, the separation of proteins, and identification of specific sets of enzymes—even the sheer inorganic chemistry—we wouldn't have been able to take even the next microstep in terms of analyzing that interaction.

That is what we are going to see over and over again. We do of course have to pay some attention to the disease aspects of parasitology, but that has been used as a way to cloak an arcane discipline that nobody but the zoologists would be able to get into, and that has had a harmful effect.

KREIER: It is interesting that molecular biology began as an effort to find out how to produce a better serum for pneumonia. There is certainly room for molecular biologists who wish to use parasitological systems as research subjects. I hope, however, that when funds get tight more conventional research is not eliminated entirely.

WARREN: Peroxide is becoming a central aspect in many different disciplines because of Klebanoff's work. Groups are now working on it in helminths, and Nogueira is working on it as a mechanism by which the host controls the parasite in Chagas' disease. Cerami has been doing interesting work on the peroxide levels inside the trypanosomes, using that in a pharmacological way as a method of killing the trypanosome. We are going to see more and more of this interaction of disciplines or of specific systems in parasitology.

BEAVER: In Mahmoud's presentation it wasn't clear whether the heavy infection was derived from the immune state or whether the immune state produced the infection. We recognize the heavy infection as the important point. What is the origin in terms of the immune state?

MAHMOUD: We don't know. If we examine the immune response of a heavily infected group of individuals—that is, the leukocyte response to specific antigens or nonspecific antigens—it is depressed only to the specific antigen.

If we try to get the monocytes in vitro we find their ability to kill the schistosome is impaired. The question is: Is it a cause or an effect? We all know the immune response is impaired in disseminated infections such as miliary tuberculosis.

We are now trying a simple experiment to treat and reexamine these patients. If they get rid of the parasite load and recover their immune responsiveness, then it was because of the parasite load. A subset of individuals in endemic areas who are susceptible to the disease do not recover their immune responsiveness.

SIMPSON: In my presentation, when I made a cryptic comment about parasitology not having a future I was being an iconoclast. I believe parasitology as a discipline will emerge with cell biology, molecular biology, and genetics as the problems become more precisely defined. I see no basic difference between the study of a parasite in a host and the study of, for example, the mitochondrion in the eukaryotic cell. It is the same biological phenomenon of metabolic and physical interdependence. Some of the basic biological problems are the interactions between cells and the interactions between cells and the environment; these problems can best be solved by the study of model systems such as parasites. I believe future molecular biologists, cell biologists, and developmental biologists will increasingly employ the parastic material provided by the natural history studies as experimental models.

WARREN: Lederberg's use of the term "new wave" is important; we have been calling it other things. The new wave itself is worthy of discussion, as is its relationship to tradition and how we can achieve the proper balance. The other important question he raised is why did it happen in bacterial infections and not in parasitic infections. This is worthy of study.

May said we have to put the beads together with a string, that is, not only do we have to look at what is going on in the laboratory in the development of new vaccines and new tools, but we have to use them in the most efficient way possible, based on models from population ecology.

TRAGER: Cerami raised an important point with regard to the economics of vaccine development. We gathered from the Nussenzweigs' presentation that there may be some candidate vaccines for malaria within a year or two. How will the Drug and Vaccine Development Corporation (DVDC) function in relation to that?

CERAMI: The idea is to have a mechanism in place in anticipation of such advances in the next few years. Part of the problem academic investigators have had is a lack of understanding of how to deal with the further development of ideas they generate in their laboratories. We are accustomed to publishing findings quickly and allowing them to enter the public domain. The problem is that research results may never be used if we don't make the necessary move to protect them and develop them commercially.

One of the things the DVDC hopes to do is simplify the process of patenting so new ideas are protected. This does not mean exploitation in a negative sense, but it is essential if new drugs or vaccines are to be developed and produced. It is not economically viable to produce a new drug or vaccine if it is not protected. Most drug companies will not consider them unless they are.

WARREN: The Rockefeller Founation made a small contribution to the Nussenzweigs' work. Several months ago they asked our advice about a genetic engineering company that was interested in their research. I asked them what progress they had made with New York University in terms of patenting the work as it went along to protect the people involved and the future utilization of their findings. They were not able to get defined, rapid, to-the-point help in this connection, so they are now negotiating with the genetic engineering company. This is regrettable because if the DVDC were functioning now it could interact profitably with scientists at a rapid level.

The other key point about the DVDC is that its proceeds will revert to the concerned individuals and institutions, and that money will be ploughed back into research on parasitic diseases and into developing an infrastructure for the pharmaceutical industries of the developing world.

LEDERBERG: I am very much behind the DVDC effort, but I can't think of a worse prototype for its exemplification than the malaria vaccine. A vaccine of any kind is an almost certain loss to the pharmaceutical industry, and one that is to be applied in developing countries is even more so.

On Cerami's question about how to anticipate the inevitable availability of candidates for vaccines, if there is anything one can be prophetic about it is that we don't want to stew around for years to solve the political-economic problems of how to proceed when we could have a headstart in terms of organizing a framework for applications. Here is a case for something to be done by government, probably through an international governmental arrangement.

We may get industrial cooperation, mostly in a philanthropic mode, because industry will attain prestige from institutional advertising. But industry should not anticipate further benefits from providing some of the material. The real expense will be in assembling clinical files and in validation, and that cannot be done without government support.

CERAMI: The DVDC is attempting to remove itself from a government background and become an independent organization that will function with less bureaucracy and fewer problems that are associated when the government carries out such activities. The idea is to see whether the precepts can be forged together.

That may be wishful thinking, but I think it can be done. The problem of making vaccines available is not only going to affect the developing world, but all Western societies. In general, because of the economics of the issue few drug companies are interested. We are going to have to phase in the DVDC with some other mechanism.

LEDERBERG: The proprietary issues are not important in this particular case; the organizational ones of trying to get the trials done predominate.

CERAMI: It is important to point out that if there were no problem and if the drug industry were doing this work there would be no need for the DVDC. The reason for creating it was to fill a gap. The idea was to have a system that would generate some activity. In ten years the developing world may be economically so affluent that people will be competing to produce new drugs and vaccines, in which case there will be no need for the DVDC.

WARREN: In the last five years 20 million Americans have traveled to areas where malaria is endemic, and most of them were not protected. There should be a huge market in this country and in Europe for a malaria vaccine.

CHERNIN: In addition to the gap Cerami mentioned, there is the parasitological gap. It has been 100 years since Patrick Manson described the transmission of "classical" filariasis by mosquitoes, yet we have barely begun to develop an animal model. It will be a long time before we get enough material for a vaccine.

Economics does not always play the kind of role Cerami described in the case of drugs. For example, pesticides play a very important part in disease control in the tropics and elsewhere. There are only four major chemical groups of pesticides, however, and all of them were developed in the early 1940s. No new pesticide group has been introduced since then, with the possible exception of the hormone pesticides, and they have not come into general use as yet. So here economics is no excuse. Industry could make a fortune by marketing the right pesticide, but is has not produced the right chemical.

CERAMI: The same thing could be said about antibiotics. The number of new ones found on the market today does not compare with the number available in the 1950s. But you have to remember that there has never been an intensive search into the parasitic diseases.

SUMMARY
OF THE CONFERENCE

Kenneth S. Warren

It is doubtful that anyone here, no matter of what persuasion, believes parasitology is in a healthy state, enjoying well-being, importance, and adequate funding, attracting the finest students, and functioning at the forefront of modern scientific methodology and theories. But it goes far beyond that. The question is not merely one of the health of parasitology, but to a considerable extent the health and well-being of people, particularly the three-quarters of mankind living in the developing world.

What is parasitology? Beaver elegantly defined the specific subject matter of parasitology. In my discussion of his paper I asserted that all infectious agents are parasites, that the field is unified by the host-parasite relationship, that subspecialization should include protozoology, helminthology, mycology, bacteriology, and virology, and that the names of departments of microbiology should be changed to departments of parasitology.

In discussing history, Schwabe spoke of the development of parasitology in the United States, particularly in terms of the integration of medicine and veterinary science. He emphasized the research-practitioner complex and the fact that these people are drawn from multiple traditions into what is basically a collaborative enterprise. Chernin made the point that in the history of parasitology emphasis has been on the plethora of parasites rather than on theory—on parasites rather than on processes.

In the teaching of parasitology, the next area we confronted, Weinstein asserted that there is no textbook available that takes processes into consideration. Current texts largely list parasites, their morphology, and their life cycles. He mentioned the Huff Report of twenty-two years ago, which discussed this restrictive view of parasitology, and stated that we need to teach more about the principles of parasitism; very little has been done about this in the last twenty years.

Weinstein reported on a 1978 survey of graduate training, which indicated that there are relatively few courses in modern immunology, biochemistry, and molecular biology. He emphasized the importance of attracting the brightest young biochemists to teaching and research in parasitology.

Burridge talked about the issues from the viewpoint of teaching veterinary parasitology. He conducted a survey of all American veterinary schools and departments, which showed, again, that there is not as much teaching as the subject warrants and that little in the course material covers modern biology.

In the area of research, Kreier discussed educational institutions. He spoke about the role of research and training in teaching institutions and the problems that arise under conditions where some believe the major emphasis should be on teaching. It should be emphasized that research does play a crucial role in teaching.

Kreier also talked about research in veterinary schools. He did a survey and found that the schools averaged three parasitologists, with a range of one to eight. This indicates it is unlikely that many of the schools have a critical mass of such scientists, and suggests they are relatively isolated. From my own experience that is even more true of medical schools.

Delappe reviewed the situation at the National Institutes of Health, dealing largely with funding of research in parasitic diseases. He presented some very important figures: in 1978 NIH support totaled about $14 million; in 1979, $16 million; and in 1980 almost $18 million. The rate of increase in funding, however, appears to be declining. In 1979 funding was increased by 16 percent, and in 1980 by 9 percent. When we consider all the pressures that are going to come to bear on the system because of the renewed interest in parasitology we must wonder about the future. Delappe concluded by describing the important Special Emphasis Programs of the NIH for the biological control of vectors and the immunology of parasitic infections.

Ruth Nussenzweig began her discussion by pointing out the need to integrate parasitology with the new biology. That seems to have been a recurring theme throughout the conference. She then talked about the radical changes in research, and the expectations that have been raised about disease control in face of the isolation of parasitology, scientifically and geographically. These are crucial issues that must be dealt with.

In the next section, training and career opportunities, we heard from two young scientists who described what attracted them to the field.

Sher, who has been a Ph.D. in immunology, was a member of a group of young people at the Salk Institute who decided to work in some field they considered to be relevant, and parasitic diseases were unanimously decided upon. Sher believes it is essential for scientists who come to parasitology from other disciplines to understand the basic biology of parasites. Another important factor is career support for young investigators who want to enter this field.

Befus, whose primary scientific discipline is parasitology, spoke about the need for flexibility, about giving oneself an opportunity to move from one aspect of the field to another, as well as to other disciplines. He made the point that the basic principles of science can be conveyed by an excellent teacher from any field of science.

With respect to sources of funding, it became clear from Cook's presentation that they are inadequate! The United States government is contributing about $49 million a year to research on parasitic diseases at the present time. Private foundation support may be important, but it is miniscule. The Clark and Rockfeller Foundations, the major philanthropic supporters of parasitology research in this country, together account for $4.7 million. The Wellcome Trust in the United Kingdom contributes another $1.3 million. Compare support for research on schistosomiasis, which amounts to 4.5 cents per person per year, with support for multiple sclerosis, $12 per year; the difference is 267 times. Cook mentioned that six drug companies alone are contributing $22 million annually to research on parasitology, an area in which they are presumably not particularly interested. If they can do that, it seems all the more ridiculous that the United States government provides such a small amount to the field.

The next section was on the present status of the literature, which can only be termed as inadequate in quantity. In 1979 nineteen major English-language serials, including all the international parasitological and tropical medicine journals, published 972 articles dealing with parasitology. In contrast, the *Journal of Immunology* alone published 960 papers. A further concern is the subject distribution: a relatively small proportion of the papers were in the newer disciplines.

As to the future of parasitology, Lederberg spoke about a "new wave," and I like the term. He referred to the great attack on bacterial and viral infections before and after World War II, which raised the hopes of the public. The present trough in the wave of public expectation is due to the failure to obtain definitive results in the subsequent attack on the great chronic diseases. Lederberg stated that it is precisely in parasitic infections that we have the analogue to the kind of opportunities for sudden and dramatic advances made by bacteriology in the past. Pointing to the shameful neglect of the health problems of the developing world, he talked about the mobilization of diverse resources with which to deal with them. He believes that with the kind of scientific nucleation represented at this conference there should be no obstacles; there are political obstacles however. We are not getting the message across that malaria is the number one disease in the world. Lederberg then addressed the broader issue of the need to treat health problems in terms of risk-benefit analysis, and the fact that even if we produce the essential drugs we run into spurious political problems that prevent us from distributing them to the people who need them. He also referred to the necessity for broad lines of communication from the laboratory level to the field level.

In his presentation, Cerami dealt with different aspects of the new wave, first biochemistry and then pharmacology. He said parasites offer a unique opportunity to do something positive about diseases, and claimed that the major concern is the need for more and better drugs. The central issue here is economics. Cerami pointed out the need for the academic profession to supplement the efforts of the pharmaceutical industry. The NIH gives no support to this area; the Walter Reed Institute, on the other hand, is carrying out research with good results. A new initiative, the not-for-profit Drug and Vaccine Development Corporation for patenting and licensing new drugs and vaccines for the diseases of the developing countries, may help in this regard.

Klebanoff discussed the interrelationship of biochemistry, pharmacology, and immunology. He stated that the patient's own body may serve as a source of pharmacological agents that are precisely delivered, using as his model the oxygen-peroxidase killing system.

Sherman stated that parasitology, pharmacology, and biochemistry generally are not effectively combined, that the approach is fragmented, and that the complexity of the field requires collaborative work. He noted the inertia that exists in attempts to reform the curriculum, and observed that greater emphasis should be placed on the physical sciences. Sherman then presented a summary of his work on the unique chemistry of the malaria parasite.

Mahmoud began by stating that the gap between parasitology and immunology is evident in the entire field of infectious diseases. In parasitic infection the host-parasite relationship has been disregarded until recently. He drew attention to the power of schistosomiasis as an immunological model in terms of immunopathology, immunoregulation, and immunity, in which a unique role for eosinophils has recently been discovered. Mahmoud made another important point about the interrelationship of field and laboratory research in regard to the types and numbers of infective organisms, the status of the host-parasite interaction, the genetically determined host responses, and the type and level of immunological functions within the host.

In discussing a protozoan parasite in this context, Nogueira observed that the standard approach to immunization, which is the induction of antibody formation, is not working, so her laboratory has developed a deep interest in the host-parasite interaction and cell-mediated immunity.

May discussed his ecological studies; many of us might tend to dismiss such work until we hear him. He talked from the point of view of populations rather than of the individual, and stated that the understanding of population ecology is crucial to the application of control measures. May made the point that laboratory investigations provide the beads, but that we need ecology to show us how best to string them together to develop cost-effective means to control diseases.

In the membrane area, the Nussenzweigs' message was simplicity of the system. This was a very important element to inject into this discussion. So often people are discouraged by the complexity; they observe the complexity and that is all they see, so they never do anything about it. Once they overcome that obstacle, they find it a lot simpler than it seemed originally.

The Nussenzweigs believe that a monoclonal antibody directed toward the single antigen on the surface of sporozoites can provide protective immunity, and that there is a possibility of using recombinant DNA in the production of enough antigen so a vaccine can be produced. This is probably one of the great breakthroughs in recent years.

Trager considered the broad issues of parasitism as a branch of parasitology and pointed out that parasitism is ubiquitous, that it occurs in all taxonomic groups and must draw on all disciplines. The main body of future work should be on the host-parasite relationship. Immunology is of course of great importance, but there are other equally fascinating aspects of the problem, such as metabolism and morphology. Trager also mentioned the opening up of the area of natural immunity, about which we know almost nothing. He noted the importance of traditional cellular parasitology, using malaria as a paradigm, and the hemoglobinopathies and the Duffy factor, as well as the importance of those who are producing new kinds of materials and new models, both in vivo and in vitro. Parasitism should be a part of every biologist's training, Trager concluded.

Simpson emphasized the importance of in vitro culture systems, calling attention to the work of H. Hirumi and G. C. Hill. In his own work Simpson uses trypanosomes as models for molecular biology. Some of the brightest young people in this country and elsewhere are moving into molecular biology. Simpson has a fine reputation in this field, and if he considers himself to be a parasitologist it is fortunate that we have parasitologists who can compete at that level. He talked about some of the practical implications of basic molecular biology. He identifies strains of trypanosomes by their minicircles and discussed what the clinical significance of this might be.

Hill made an exceedingly important point when he emphasized the way complex biochemistry can enable the parasite to escape the effects of pharmacological agents.

Finally, Gross discussed parasitology from the point of view of a developmental biologist who is not a parasitologist. He emphasized the problem of the control of parasites, which is what convinced the Rockefeller Foundation and the MBL to enter this field. Another point Gross brought up, which is of immense significance, is the interface of cell and molecular biology with ecology. The MBL summer courses began in 1888, and although parasitology was not one of them, many eminent parasitologists have come there to conduct research each summer. Gross said the MBL is a microcosm of what will happen elsewhere if we want it to happen. If the parasitism course succeeds, the whole concept may be catalytic for the development of the new wave of parasitology and of host-parasite relationships in departments of biology, and even of medicine, in the United States. Major biological issues get thrashed out quickly in this pressure cooker at Woods Hole. Gross also mentioned the significance of laboratory teaching, which that course very clearly revealed. Another point Gross made is that we have to get over the idea that molecular biology and immunology are arcane subjects. Once people use these techniques the field turns out not to be as arcane, specialized, and expensive as they had believed. The participants not only learned about hybridomas and recombinant DNA; they learned a great deal about parasites. It is important to note that the confluence of the two strains do not cause any great amount of pain.

I would like to close with some issues we might take as a kind of consensus or opportunity.

Outsiders such as Sher, Cerami, and Klebanoff are moving

into the field because they consider it a great opportunity. Lederberg sees it as providing for sudden and dramatic advances. Whereas investigators studying cancer are running into a wall and getting nowhere, we suddenly see in parasitology the opportunity to make breakthroughs in an accessible field that has been in a relatively dormant state.

With respect to the status of the field, there is a general consensus that we have to do some proselytizing about the inadequate funding with respect to the potential of parasitology in dealing with four subjects—biochemistry, pharmacology, immunology, and membranes. All these will combine to give us a greater understanding of host-parasite relationships, to produce new drugs and vaccines, and to provide proper and cost-effective applications.

One of the great problems here is how to preserve and foster the best of the traditional along with the new, how to achieve collaboration and synergy. It may take some time, however, to get the traditional departments of biology moving in these directions.

There are opportunities for some key institutions to lead the way. One of these is the Rockefeller University; another is the Massachusetts Institute of Technology, which is considering developing a major program in the molecular biology of parasitism. The MBL has already begun, and I believe there is a possibility that the biology of parasitism may become a major, year-round program in Woods Hole.

Our greatest concern at this point is that funding agencies have not yet realized the opportunities, so we must do everything in our power to increase their awareness, not for the sake of parasitology, alone, but for the health of people throughout the world.

PARTICIPANTS

Paul C. Beaver, Ph.D.
Emeritus William Vincent
 Professor of Tropical Diseases
 and Hygiene
Department of Tropical Medicine
School of Public Health and
 Tropical Medicine
Tulane University
New Orleans, Louisiana

A. Dean Befus, Ph.D.
Assistant Professor
Department of Pathology
School of Medicine
McMaster University
Hamilton, Ontario, Canada

John Z. Bowers, M.D.
New York, New York

Michael J. Burridge,
 B.V.M. & S., Ph.D.
Associate Professor
Department of Epidemiology
Head
Division of Tropical Animal Health
Department of Preventive Medicine
College of Veterinary Medicine
University of Florida
Gainesville, Florida

Anthony Cerami, Ph.D.
Professor and Head
Laboratory of Medical Biochemistry
The Rockefeller University
New York, New York

Eli Chernin, Sc.D.
Professor
Department of Tropical Health
School of Public Health
Harvard University
Boston, Massachusetts

Joseph A. Cook, M.D.
Program Officer
The Edna McConnell Clark
 Foundation
New York, New York

Irving P. Delappe, Ph.D.
Chief
Molecular Microbiology and
 Parasitology Branch
National Institute of Allergy
 and Infectious Diseases
National Institutes of Health
Bethesda, Maryland

Paul R. Gross, Ph.D.
President and Director
Marine Biological Laboratory
Woods Hole, Massachusetts

George C. Hill, Ph.D.
Associate Professor
Department of Pathology
Colorado State University
Fort Collins, Colorado

Seymour J. Klebanoff, M.D.
Professor
Department of Medicine
School of Medicine
University of Washington
Seattle, Washington

Julius P. Kreier, V.M.D., Ph.D.
Professor
Department of Microbiology
College of Biological Sciences
The Ohio State University
Columbus, Ohio

Joshua Lederberg, Ph.D.
President
The Rockefeller University
New York, New York

Adel A. F. Mahmoud, M.D., Ph.D.
Professor of Medicine
Director
Division of Geographic Medicine
Department of Medicine
School of Medicine
Case Western Reserve University
Cleveland, Ohio

Robert M. May, Ph.D.
Professor
Department of Biology
Princeton University
Princeton, New Jersey

Nadia Nogueira, M.D., Ph.D.
Assistant Professor
Laboratory of Cellular Physiology
  and Immunology
The Rockefeller University
New York, New York

Ruth S. Nussenzweig, M.D., Ph.D.
Professor and Head
Division of Parasitology
Department of Microbiology
School of Medicine
New York University
New York, New York

Victor Nussenzweig, M.D.
Professor
Department of Pathology
School of Medicine
New York University
New York, New York

John A. Pino, Ph.D.
Director
Agricultural Sciences
The Rockefeller Foundation
New York, New York

Calvin W. Schwabe, D.V.M., M.P.H.
Professor
Department of Epidemiology
  and Preventive Medicine
School of Medicine
University of California, Davis
Davis, California

F. Alan Sher, Ph.D.
Assistant Professor of Pathology
Division of Parasite Immunology
Department of Medicine
Harvard Medical School
Boston, Massachusetts

Irwin W. Sherman, Ph.D.
Professor of Zoology
Department of Biology
University of California, Riverside
Riverside, California

Larry Simpson, Ph.D.
Professor
Department of Biology
University of California, Los
  Angeles
Los Angeles, California

William Trager, Ph.D.
Professor
Laboratory of Parasitology
The Rockefeller University
New York, New York

Kenneth S. Warren, M.D.
Director
Health Sciences
The Rockefeller Foundation
New York, New York

Paul P. Weinstein, Sc.D.
Professor
Department of Biology
University of Notre Dame
Notre Dame, Indiana

INDEX